Unsupervised and Semi-Supervised Learning

Series Editor

M. Emre Celebi, Computer Science Department, Conway, AR, USA

Springer's Unsupervised and Semi-Supervised Learning book series covers the latest theoretical and practical developments in unsupervised and semi-supervised learning. Titles – including monographs, contributed works, professional books, and textbooks – tackle various issues surrounding the proliferation of massive amounts of unlabeled data in many application domains and how unsupervised learning algorithms can automatically discover interesting and useful patterns in such data. The books discuss how these algorithms have found numerous applications including pattern recognition, market basket analysis, web mining, social network analysis, information retrieval, recommender systems, market research, intrusion detection, and fraud detection. Books also discuss semi-supervised algorithms, which can make use of both labeled and unlabeled data and can be useful in application domains where unlabeled data is abundant, yet it is possible to obtain a small amount of labeled data.

Topics of interest in include:

- Unsupervised/Semi-Supervised Discretization
- Unsupervised/Semi-Supervised Feature Extraction
- Unsupervised/Semi-Supervised Feature Selection
- Association Rule Learning
- Semi-Supervised Classification
- Semi-Supervised Regression
- Unsupervised/Semi-Supervised Clustering
- Unsupervised/Semi-Supervised Anomaly/Novelty/Outlier Detection
- Evaluation of Unsupervised/Semi-Supervised Learning Algorithms
- Applications of Unsupervised/Semi-Supervised Learning

While the series focuses on unsupervised and semi-supervised learning, outstanding contributions in the field of supervised learning will also be considered. The intended audience includes students, researchers, and practitioners.

** Indexing: The books of this series indexed in zbMATH **

Ibraheem Alharbi • Chiheb-Eddine Ben Ncir •
Bader Alyoubi • Hajer Ben-Romdhane
Editors

Advances in Computational Logistics and Supply Chain Analytics

Springer

Editors
Ibraheem Alharbi
College of Business
University of Jeddah
Jeddah, Saudi Arabia

Chiheb-Eddine Ben Ncir
College of Business
University of Jeddah
Jeddah, Saudi Arabia

Bader Alyoubi
College of Business
University of Jeddah
Jeddah, Saudi Arabia

Hajer Ben-Romdhane
Institut supérieur de Gestion (ISG)
University of Tunis
Le Bardo, Tunisia

ISSN 2522-848X ISSN 2522-8498 (electronic)
Unsupervised and Semi-Supervised Learning
ISBN 978-3-031-50035-0 ISBN 978-3-031-50036-7 (eBook)
https://doi.org/10.1007/978-3-031-50036-7

This Springer imprint is published by the registered company Springer Nature Switzerland AG
The registered company address is: Gewerbestrasse 11, 6330 Cham, Switzerland

Paper in this product is recyclable.

Preface

This volume provides advances in computational logistics and supply chain analytics. It includes innovative data-driven and learning-based approaches, methods, algorithms, techniques, and tools that have been designed or applied to create and implement a successful logistics and supply chain management process. Such computational techniques and methods allow companies to better meet consumer expectations, to deliver products faster, and to largely reduce costs. The volume provides a systematic understanding of the scope in-depth and rapidly builds an overview of recent computational methods and techniques applied to ensure continuous improvement of transport, logistic, and supply chain management processes.

The volume opens with chapter entitled "The Dynamic Vehicle Routing Problem: A Comprehensive Survey". In this chapter, Nasreddine Ouertani, Hajer Ben Romdhane, and Saoussen Krichen provide an in-depth review of the dynamic vehicle routing problem (DVRP) within the context of supply chain management. The authors conducted a survey over more than 131 articles published from 2017 until the first quarter of 2023 and explore the main solutions proposed to solve the DVRP problem. The authors also highlighted real-world applications and discussed future trends and directions related to DVRP.

The vehicle routing problem is also investigated in chapter entitled "Multi-objective Optimization for Electric Vehicle Routing Problem: A Literature Review". Given the growing popularity of electric vehicles and the need to reduce greenhouse gas emissions, Anouar Haddad, Takwa Tlili, and Saoussen Krichen focused on solving the electric vehicle routing problem (EVRP) that involves finding optimal routes for electric vehicles to minimize energy consumption while satisfying various constraints such as battery capacity, charging station availability, and time windows. The authors provide a comprehensive literature review of EVRP and its variants, as well as a variety of solution algorithms and applications of EVRP in different domains, such as transportation and logistics.

The next chapter also deals with a specific case of vehicle routing problem that involves finding the shortest routes for a group of identical vehicles stationed at a central location with the objective of delivering goods to a set of customers within a specific time frame. Takwa Tlili, Marwa Harzi, and Saoussen Krichen in chapter, "A Decision Support System for Solving the Windy Rural Postman Problem," address this challenge by considering a double time window where the first represents travel time, and the second represents the time for postal delivery. The authors propose a variable neighborhood-based approach and demonstrate its efficiency through real-life instances of a postal company in northwest Tunisia.

In chapter entitled "A Hybrid Meta-Heuristic to Solve Flexible Job Shop Scheduling Problem," Makram Zaidi, Amina Amirat, Bassem Jarboui, and Abdelkrim Yahyaoui address the flexible job shop scheduling problem (FJSP). The authors propose an integrated meta-heuristic approach based on both hybrid genetic algorithm and simulated annealing. The proposed approach incorporates an acceptance criterion that enables maintaining high characteristics of the previous generations and allows the reduction of the disruptive effects of genetic operators. The authors demonstrate the effectiveness of the proposed approach in achieving better performance within a reasonable computational time.

The use of meta-heuristics and evolutionary algorithms for solving logistic problems is also investigated in chapter under the title "Multimodal Freight Transport Optimization Based on Economic and Ecological Constraint". Lilia Rejeb, Abir Chaabani, Hajer Safi, and Lamjed Ben Said address the problem of optimizing multimodal freight transport by considering additional economic and ecological objectives rather than the typical focus on costs. Three distinct objectives have been simultaneously considered: overall transportation cost, transportation time, and CO_2 emissions. The authors effectively solved such a challenging optimization problem by using the tabu search and the genetic algorithms and demonstrated the effectiveness of the proposed approach through a real-world experimental study.

Rather than considering a simultaneous optimization of several objectives, Abir Chaabani and Lamjed Ben Said propose in chapter to hierarchically address the problem of optimizing the decisions between the production and the distribution entities under the constraints of shared depot resources in the distribution phase. The chapter entitled "Solving Hierarchical Production-Distribution Problem Based on MDVRP Under Flexibility Depot Resources in Supply Chain Management" proposes a new formulation of the problem based on a mixed bi-level framework allowing to adopt a hierarchical decision-making process where lower-level decisions depend on upper-level actions. The mixed bi-level problem is solved through an intelligent and effective cooperative decomposition-based algorithm. The authors demonstrate the competitive performance of the proposed algorithm in reducing the total traveling costs of generated solutions and highlight the benefits of the flexible choice of stop depot.

The last two chapters deal with logistics in the Kingdom of Saudi Arabia by studying inhibitor factors and financial performance measurement for a successful implementation of a digital supply chain by using computational methods. In chapter entitled "Analysis of Inhibitors to Implementing Digital Supply Chain in Saudi Arabia: An Interpretive Structural Modelling (ISM) Approach," Raouf Jaziri, Abdullah Alshareef, Saleh Alnahdi, and Mohammad Miralam focus on identifying and analyzing the main inhibitors of digital supply chain adoption in the Saudi

Arabia. The authors propose a two-phase approach. In the first phase, several structural variables are identified based on existing literature and expert opinions, while in the second phase, an interpretive structural modeling (ISM) based on the matrix of cross-impact multiplications is adopted to classify inhibitors according to their influence, dependence, and driving power. The findings provide valuable insights for Saudi policymakers and logistics managers in effectively deploying a digital supply chain. Chapter entitled "Financial Performance Measurement of Logistics Companies: Empirical Evidence from Saudi Arabia" examines the main performance measures affecting the financial situation of logistics companies listed on the Saudi stock exchange over the period 2018–2021. Raef Bahrini, Ahmed Zamzam, and Assaf Filfilan adopted a multi-criteria decision-making approach to identify the best performance measures affecting the financial performance of these companies. The findings highlight profitability as one of the principal performance indicators affecting the financial performance of Saudi logistics companies followed by liquidity and solvency ratios, respectively.

We hope that the proposed volume ensures a holistic view of computational logistics and supply chain management and offers valuable insights, literature reviews, and innovative methodologies to address these challenges. We hope that all covered topics, from dynamic vehicle routing to digital supply chain implementation, will help professionals, researchers, and students in logistics, transportation, and supply chain management to better understand and solve complex logistic problems.

Jeddah, Saudi Arabia — Ibraheem Alharbi
Jeddah, Saudi Arabia — Chiheb-Eddine Ben Ncir
Jeddah, Saudi Arabia — Bader Alyoubi
Le Bardo, Tunisia — Hajer Ben-Romdhane

Acknowledgements

The editors extend their appreciation to the Deputyship for Research & Innovation, Ministry of Education in Saudi Arabia for funding this research work through the project number MoE-IF-UJ-22-Book-1.

Contents

About the Editors

Ibraheem Alharbi is currently the dean of the College of Business at the University of Jeddah. He received a BA degree from King Abdul Aziz University, Saudi Arabia, in 2002, and a Master's and PhD degree from La Trobe University, Australia, in 2009. He serves as full professor in the Department of Management Information Systems, College of Business, University of Jeddah, Jeddah, Saudi Arabia. His research interests include business and information ethics, information privacy, and electronic commerce. He has published many research articles in reputed journals and participated in many international conferences.

Chiheb-Eddine Ben Ncir is currently associate professor at the University of Jeddah, Saudi Arabia, and member of LARODEC laboratory (University of Tunis). He received his PhD in computer science and management from Higher Institute of Management, University of Tunis, in 2014 and a HDR degree (Habilitation for the Supervision of Doctoral Research) in 2021. He previously occupied the position of assistant professor at the Higher School of Digital Economy (University of Manouba, Tunisia) from 2015 to 2018. His research interests concern machine learning methods and data mining tools with a special emphasis on big data clustering, disjoint and non-disjoint partitioning, kernel methods, as well as many other related fields. He is the author or coauthor of more than 50 publications in several prestigious journals and conferences. He is a regular reviewer for many refereed international journals and is a coeditor of some Springer books.

Bader Alyoubi currently serves as the dean of the College of Sports Sciences at the University of Jeddah. Additionally, he holds the position of professor of Management Information Systems at the College of Business in the same university. His research focuses on decision support systems and knowledge management techniques, and their applications in various sectors such as business, government, health, and decision-making. He has authored and coauthored more than 50 publications in his field of specialization. Moreover, he is the founder of the Saudi Center for Preparing and Empowering Entrepreneurs at the University of Jeddah.

Bader Alyoubi has made significant contributions to the establishment of colleges and scientific disciplines at the University of Jeddah, and he also serves as the head of multiple committees.

Hajer Ben-Romdhane is an assistant professor in the Department of Computer Science at Institut Supérieur de Gestion, University of Tunis, Tunisia, and member of LARODC laboratory. She received a PhD degree in Business Computer Science from the University of Tunis in 2016. She is currently head of the Department of Distance Learning at the University of Tunis. Her research interests include modeling of complex and dynamic optimization problems, multiple criteria decision-making, and the design of decision support systems/recommender systems for transportation, operations management, and health care. She has authored and coauthored many refereed journal and conference papers. She guest-edited several special issues and has served as publication chair of several conferences held in Tunisia in the few last years.

The Dynamic Vehicle Routing Problem: A Comprehensive Survey

Nasreddine Ouertani, Hajer Ben-Romdhane, and Saoussen Krichen

1 Introduction

Supply chain management (SCM) represents the process of planning, organizing, and managing the flow of goods or services and information from raw materials to finished products, and from the point of origin to the point of consumption. It involves various activities, including procurement, production planning, inventory management, logistics, and customer service. The goal of SCM is to optimize the entire process, from start to finish, to ensure that products are delivered to customers quickly, efficiently, and at the lowest possible cost. It also helps companies to reduce costs, improve profitability, and mitigate risks associated with supply chain disruptions. The COVID-19 pandemic has highlighted the importance of SCM, as disruptions to global supply chains have caused significant disruptions to businesses and economies worldwide [2]. Companies that have effective SCM practices in place have been better equipped to respond to these disruptions and mitigate their impact.

Transportation undoubtedly represents an important aspect of SCM that can be defined as the movement of products from one location to another, from suppliers to manufacturers, wholesalers to final customers. Transportation is a critical component of the supply chain as it can impact delivery times, product

N. Ouertani (✉)
Université de Tunis, Institut Supérieur de Gestion, LARODEC Laboratory, Tunis, Tunisia

LASPI, Université Jean Monnet, IUT de Roanne, Saint-Étienne, France
e-mail: nasreddine.ouertani@gmail.com

H. Ben-Romdhane · S. Krichen
Université de Tunis, Institut Supérieur de Gestion, LARODEC Laboratory, Tunis, Tunisia
e-mail: hajer.benromdhan@isg.u-tunis.tn; saoussen.krichen@isg.rnu.tn

I. Alharbi et al. (eds.), *Advances in Computational Logistics and Supply Chain Analytics*, Unsupervised and Semi-Supervised Learning,
https://doi.org/10.1007/978-3-031-50036-7_1

quality, and customer satisfaction. In operations management and decision-making, much attention is devoted to the delivery process in the supply chain, which is widely studied as one of the most critical and extensively researched activities [69]. Thus, the process of designing delivery (or pickup) routes to service customers is one of the central problems of SCM, and it is often performed by a fleet of vehicles known in the literature as Vehicle Routing Problem (VRP). Indeed, this problem was introduced at the first time in [32] as "Truck Dispatching Problem." The objective of the problem was to create a model that could efficiently allocate a fleet of identical trucks to meet the demand for oil from various gas stations, all while minimizing the total travel distance. Later, in 1964, [31] generalized this problem and introduced it as the VRP, which focused on supplying a set of geographically dispersed customers from a central depot. Each customer has a known location and a non-negative demand. In addition, the routing task is carried out using a fleet of homogeneous vehicles departing and arriving at the depot. The main objective of VRP is to satisfy all customers' demands with the lowest travel distance (or cost). However, in the majority of cases, this objective alone does not touch the reality and respond to the real needs of companies. Over the years, extensive research has focused on integrating various factors into the VRP, including customer satisfaction, vehicle fleet management, and environmental considerations. As a result, numerous extensions to the classical version of the VRP have been proposed, each with unique features that may affect either the objective function or the set of constraints [130]. The VRP literature has extensively focused on the static version of the problem over the years [6]. In the static VRP, all problem features are certain before the start of the working day, and no changes are expected to occur.

However, the dynamic nature of real-life situations presents challenges when attempting to model certain real-world applications, including taxi services, courier services, and repair services [102, 105, 110]. Traditional routing problem models may not be suitable for these applications due to the changing nature of the data and information involved. In such situations, it is crucial for decision-makers to take into account newly revealed information, including factors such as customer locations, customer demands, service times, and travel times for vehicles. This dynamic information needs to be incorporated into the planning process, allowing for the re-adaptation of routes to ensure efficient and effective deliveries [105].

In real-world scenarios, decision-makers are not only focused on achieving an economic gain but also on a solution that can adapt the routing plan when unexpected events appear such as vehicle breakdowns, variation in the number of customers to be visited, and route closures due to roadworks or accidents. Undoubtedly, the routing problem becomes significantly more complex and challenging when dealing with dynamic situations. Thus, this class of problem is named as: dynamic VRP (DVRP). Indeed, in DVRP, the available information can be revealed or updated over time and should be treated in such a way the planned route must be adjusted to respond to all new variations [72, 104, 142]. Clearly, to better deal with the problem, a real-time communication between the dispatcher (decision-maker) and the drivers until the end of the delivery process is required. Thanks to the recent advances in information and communication technologies, companies became able to manage vehicle routes in real time [105]. Due to the ability of modeling real-world situations, an increasing interest has focused on DVRP and its solution methods [110]. Furthermore, DVRPs have been used to model many applications in different domains such as: repair services, emergency services (police, fire, and ambulance services), and dial-a-ride systems [71, 104, 105].

Our survey seeks to offer a thorough and up-to-date examination of the current research on the DVRP at the aim to provide readers with an understanding of the current state of research in this field, as well as highlight potential areas for future investigation. To achieve this, we have surveyed a total of 131 articles published from 2017 until the first quarter of 2023, which cover various aspects of the DVRP, including its main variants, features, limitations, and solution methods. By analyzing these articles, we present a new and extensive categorization, with a particular emphasis on identifying gaps in the research and highlighting opportunities for future investigation in the DVRP, specifically for conventional, electric, and hybrid vehicles.

This chapter is intended to provide a comprehensive and an updated overview of the DVRP and its specific features. This chapter addresses the need to update the literature on DVRP, since the number of published articles in this field has significantly increased over the past seven years. Indeed, the authors in [109] reported the publication of 80 articles between 2015 and 2021, while [105] reviewed 117 published articles on DVRP up to 2014. Furthermore, besides the dynamic nature of real-life applications, also, they involve the optimization, concurrently, of many conflicting objectives [78]. In response to the dynamic multi-objectives changes in problem features of real-world applications over time, we have introduced a new classification system based on sets of objective functions. This classification allows for a comprehensive and effective approach to tackling the evolving problem features and objectives over time. Finally, we extended the previous taxonomy by introducing a new classification that focuses on the benchmarks and performance measures used in DVRP.

The rest of this chapter is organized as follows. Section 2 is dedicated to present the methodology of our survey. Section 3 provides a brief discussion about VRP, including the mathematical formulation, main important methods for solving VRP, and some important extensions of this problem. Section 4 is devoted to the presentation of DVRP, the dynamism measure of such a problem, a comparison between static and dynamic routing problems. Section 5 highlights the main variants of the dynamic problem. Section 6 is focused on the presentation of the dynamic multi-objective optimization problem. Section 7 is dedicated to elucidate the main existing solution methods dealing with the DVRP. Section 8 is dedicated to present the most investigated performance measures employed to evaluate solution approaches. Section 9 is devoted to present the main used benchmarks. Section 10 presents the main existing applications of DVRPs. Section 11 provides a summary of significant trends and future perspectives for the development of research into the DVRP. Finally, Sect. 12 concludes the chapter.

2 Survey Methodology

To conduct our survey, we searched for articles published in scholarly, academic journals related to the DVRP using the following keywords: "Dynamic," "Real Time," and "On-line." Thus, we have identified 131 articles that met our search criteria. We analyzed and synthesized the information from these articles to gain a better understanding of the current state of research in this area.

Figure 1 illustrates the number of articles published in different journals, mainly from 2017 to the first quarter of 2023, providing an overview of the publication trends in the field of the DVRP. Indeed, the analysis highlights the prominence of the European Journal of Operational Research, Computer & Operations Research, and Swarm and Evolutionary Computation as the preferred journals for DVRP authors. These three journals account approximately 47% of all DVRP published articles during the specified period. This observation indicates the significant influence and recognition of these journals within the DVRP research community.

To better analyze the surveyed articles, we have used VOSviewer [134], which is a software tool for scientometric analysis that enables the creation and visualization of bibliometric networks of scholarly articles. Additionally, this software allows to better analyze the structure and patterns of scientific literature and identify key authors, publications, and research topics. Figure 2, generated using VOSviewer, provides a visual representation of the main variants, solution methods, and real-life applications of DVRP based on a scientometric analysis of surveyed articles. In this figure, nodes represent different solution methods for the DVRP, such as

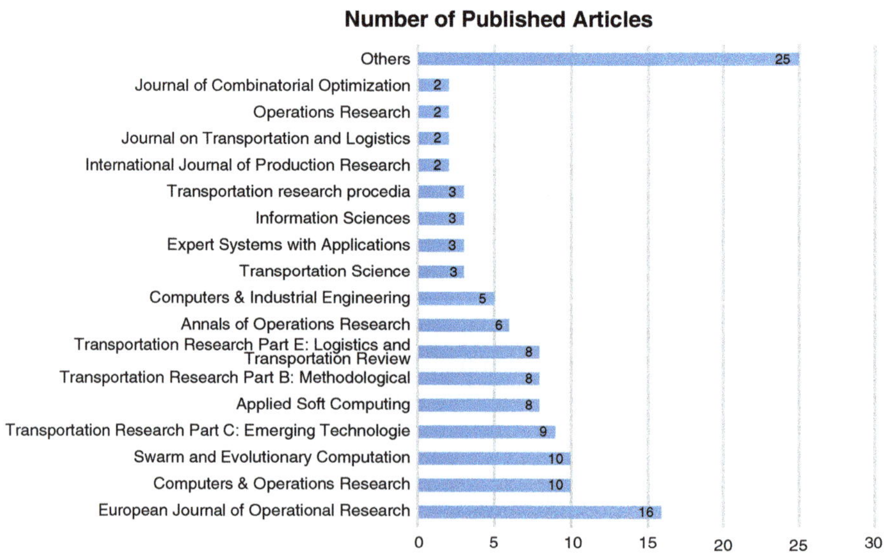

Fig. 1 Published articles per journals

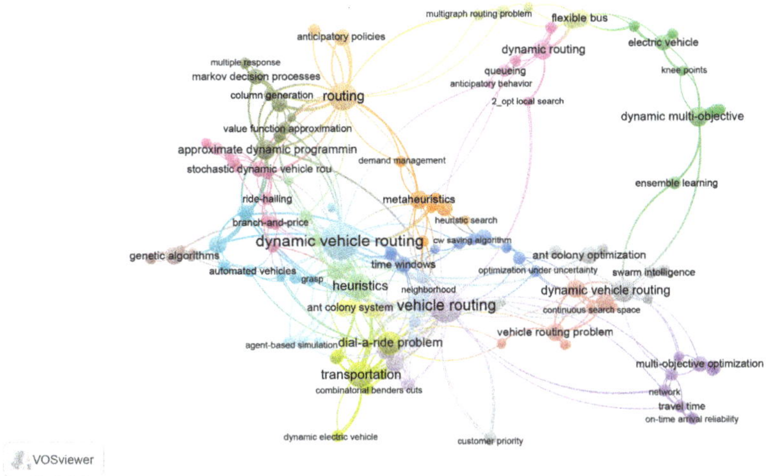

Fig. 2 Visualizing key variants, solution methods, and applications of the DVRP

Markov Decision Process (MDP), Column Generation (CG), and Genetic Algorithm (GA). These nodes are grouped together based on similarities in the optimization techniques or problem formulations they use. Furthermore, other nodes depict several real-world applications of the DVRP, such as electric vehicles (EVs) and flexible bus. On the basis of this figure, we can clearly identify the relationships between the different DVRP applications and the solution methods studied in the literature. Also, the different nature of the problem, such as the multi-objective and the continuous optimization, the uncertain modeling of the problem.

3 Presentation of the Vehicle Routing Problem

This section is dedicated to present the mathematical formulation of the VRP and the main existing variants.

3.1 Problem Statement

Formally speaking, VRP can be mathematically modeled by using an undirected complete graph $G = (V, E)$ with $V = \{0, 1, \ldots, N\}$ the set of vertices and E the set of edges. The vertex 0 represents the depot, whereas the other vertices represent the customers. Each customer has spatial coordinates X_{coord} and Y_{coord}. Let N represent the set of customers and q_i denote the demand of customer i. The capacity of a vehicle is limited to Q, which must be respected. The distance between

Table 1 The parameters and descriptions of the VRP

Parameters	Description
N	Set of customers
K	Number of vehicles, where $k \in \{1, .., K\}$
Q	Capacity of vehicle
M	Large number
q_i	Demand of customer i, where $i \in \{1, .., N\}$
d_{ij}	Transportation distance from customer i to j, where $i \neq j$

Table 2 The decision variables of the VRP

$$x_{ijk} = \begin{cases} 1 & \text{if vehicle } k \text{ travels from customer } i \text{ to } j. \\ 0 & \text{otherwise.} \end{cases}$$

$$y_{ik} = \begin{cases} 1 & \text{if vehicle } k \text{ visits customer } i. \\ 0 & \text{otherwise.} \end{cases}$$

customers i and j, where $i, j \in N$ and $i \neq j$, is denoted by d_{ij}. The transportation distance is calculated using the Euclidean distance. In what follows, Tables 1 and 2 present the set of used parameters and the two decision variables.

The mathematical model of VRP can be expressed as a mixed integer linear programming (MILP) problem as follows [3]:

$$Min \ \ Z(x) = \sum_{k=1}^{K} \sum_{i=0}^{N} \sum_{j=0}^{N} d_{ij} x_{ijk} \tag{1}$$

Subject to

$$\sum_{i=1}^{N} y_{ik} q_i \leq Q \ \ \forall k \in K \tag{2}$$

$$\sum_{k=1}^{K} y_{ik} = 1 \ \ \forall i \in N \tag{3}$$

$$\sum_{k=1}^{K} y_{ik} = K \ \ for \ \ i = 0 \tag{4}$$

$$\sum_{i=0}^{N} x_{ijk} = y_{jk} \ \ \forall j \in [0, .., N], \ \ \forall k \in K, \ \ i \neq j \tag{5}$$

$$\sum_{j=0}^{N} x_{ijk} = y_{ik} \quad \forall i \in [0, .., N], \quad \forall k \in K, \quad i \neq j \tag{6}$$

$$\sum_{i,j \in S} x_{ijk} \leq | S | - 1 \qquad S \subset N, \quad | S | \geq 2, \quad \forall k \in K, \quad i \neq j \tag{7}$$

$$x_{ijk}, y_{ik} \in \{0, 1\} \quad \forall i, j \in [0, .., N], \quad \forall k \in K, \quad i \neq j \tag{8}$$

The objective function of VRP is represented in Eq. 2, which aims to minimize the total travel distance. The set of constraints from 2 to 8 is defined as follows: Constraints 2 ensure that the vehicle capacity is not exceeded, while Constraints 3 ensure that each customer is visited only once by a single vehicle. Constraints 4 state that every vehicle starts from the depot. Constraints 5 and 6 establish the relationship between the decision variables. Constraints 7 eliminate sub-tours, and Constraints 8 enforce the binary restriction of decision variables.

3.2 *Main Variants of the VRP*

VRP is a well-studied problem in the operations research literature [15]. Indeed, since it was proposed in 1959, this problem has modeled several real-life applications and many extensions have been introduced. According to [16], each year, the rate of published works dealing with VRP is increasing exponentially by 6%. Hence, many books and surveys dealing with VRP have been published [16, 35].

Further, a lot of effort have been devoted to model real-world situations and give a raise to several VRP variants. In fact, from one variant to another, the objective can be modified such as the minimization of the transportation risk [98], the minimization of CO_2 emissions [99], and the maximization of customer satisfaction [145]. Furthermore, additional constraints can be added. Indeed, this may include vehicle features such as: the types of operating vehicles (homogeneous/heterogeneous) [100] with single or multi-compartment [96] and whether the number of vehicles is fixed or not [119]. Moreover, there are other practical aspects concerning the types of services provided to customers as delivery and/or pickup [68]. Besides, considering the customer preferred time windows as constrained to be respected [81]. Additionally, the service can be performed over several days [20]. In addition, the type of the transported materials can add up a supplementary complexity to the original problem: dangerous or delicate materials that should be transported with care [98]. Furthermore, in some situations, additional constraints related to the depot may be considered such as: products are picked up from several depots (multi-depot) [146] and vehicles are not required to return to the depot (open VRP) after finishing their servicing [17]. Also, information in real-world applications are subject to change during the planning horizon and revealed over time [97]. Moreover, it should be noted that many existing works have combined several variants of VRP such as: serving customers within a predefined time using a fleet of heterogeneous [91]. A multi-compartment VRP with time windows is used in [23] to satisfy customer's need. Moreover, time-dependent DVRP is investigated in the work of [51], where the travel time between two customers is considered variable through the time and customer demands appeared over the planning horizon. In addition, several objectives have to be optimized simultaneously [145], and many others.

4 Dynamic Vehicle Routing Problem

We start this section by presenting the DVRP and its formal description. Next, we explain the concept of degree of dynamism. In order to better identify the main feature of DVRP, we highlight the differences between the static and dynamic variants of the problem.

4.1 Problem Description

Thanks to the advancement of communication and information technology, decision-makers are more aware of the importance of just-in-time managerial strategies [98]. It gives rise to a new class of problems, namely DVRP also known as real-time or online VRP, where new demand appeared as time progresses and the route must be updated dynamically to integrate these new demands into the routing process [88, 102, 105, 110]. As stated in [104], "a problem is dynamic when some part of the input data is revealed to the solver during optimization." This means that we cannot build a fixed solution beforehand, and we have to adjust the solution whenever the problem changes. To the best of our knowledge, the first attempt to consider real-time requests in the context of DVRP can be found in [59]. In that study, a sequential insertion procedure was used to deal with new requests as they arose in real time. In contrast with static VRP where all problem features (e.g., geographical location of the customers, the number of clients, customer requests, service time) are known with certainty in advance (prior to the day starts) and do not change after the routes have been planned (deterministic information), in DVRP some or all of the inputs are subject to change during the planning horizon such as: some routes were blocked due to bad weather conditions or traffic jam that can lead to increase the travel time and new and/or canceling requests. Therefore, in DVRP the planner (decision-maker) proceeds as follows. First, an initial routing plan is created including all known customers in advance (available information). After that, when a new request occurs (environment changes), the online requests have to be incorporated into the scheduling plan with the remaining customers (customer not yet served) while respecting the system constraints (e.g., vehicle capacity, service time). In cases where there is no feasible insertion, the new customer is assigned to an unused vehicle if there is one available at the depot. Otherwise, the

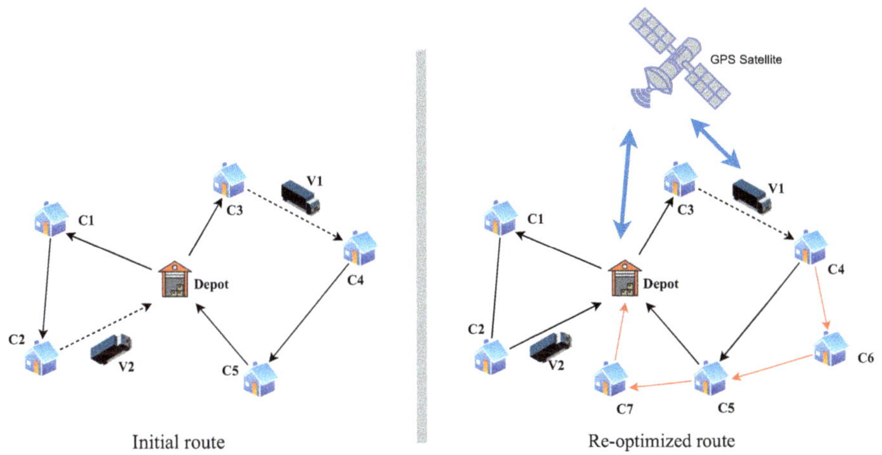

Fig. 3 Graphical depiction of the DVRP

request is rejected. Since the initial plan has re-adapted, the dispatcher has to inform the vehicle to the next destination using the Global Positioning System (GPS), for example.

As we stated above, the dynamic version of VRP requires real-time communication between the dispatcher (decision-maker) and the drivers until the end of the delivery process [102]. Mainly, two equipment are needed to operate vehicles in a real time. The first one is the *positioning equipment* such as GPS, geographic information system (GIS). And a *communication equipment* is essential to inform the drivers about the new destination as: smart phone, text message.

Clearly, DVRP deals with two types of requests: *advance* and *immediate* requests [72]. The first one are used to design the preplanned route, while the immediate requests are revealed to the planner over time and have to be incorporated to the routing plan to meet customer needs [142].

In Fig. 3, an example of DVRP is illustrated. The left side of the figure shows the routing plan generated initially, where solid arrows indicate the order of visiting the known customers and dashed arrows indicate the current route. On the right side of the figure, new online demands are received, resulting in the addition of new locations to the map. Assuming that these demands are received, while vehicle 2 (V_2) is traveling across the arc between customers C3 and C4, the rest of the trajectory of V_2 is adapted to include customers C6 and C7, starting from C4. This adaptation is represented by the red arrows connecting the customers' locations in the figure.

It is worth noting that in DVRP, the entire service process is carried out within a single period, which typically lasts for a day. This involves serving a set of known and unknown customers (in advance). This is in contrast to periodic VRP (PVRP) [20], which is aimed at serving a set of known customers over one or more days (multi-period). For example, a PVRP for hazardous waste management is presented in [106], where the authors proposed a model for collecting and transporting industrial hazardous waste from known sources to recycling facilities over a multi-period horizon.

4.2 The Degree of Dynamism

An important concept related to DVRP is the degree of dynamism (*dod*). This measure models the number of dynamic events or how much the problem is dynamic [83]. Two major factors contribute to the *dod* of a VRP with dynamic requests: the number of advance requests n_s and the number of immediate requests n_d. Therefore, the *dod* formula is given by

$$dod = \frac{n_d}{n_s + n_d}. \tag{9}$$

The *dod* values take values between 0 and 1, where 0 indicates that no requests are received during the service period and 1 means that all requests are received in real time. For instance, if the *dod* value is 0.5, then 50% of the customer demands are known in advance and are used to create the initial routing plan. The remaining customers form the dynamic demands that are revealed during the course of the service period.

Later, authors in [71] generalized the *dod* by adding dynamic appeared time and customer time window, and he proposed the *effective degree of dynamism* indicated by *edod*. The *edod* is calculated as follows:

$$edod = \frac{\sum_{i=1}^{n_d} (\frac{t_i}{T})}{n_s + n_d}, \tag{10}$$

where t_i denotes the appeared time of the ith customer during the planning horizon $[0, T]$. Therefore, the *edod* represents an average of how late the requests are occurred compared to the latest possible time the requests could be received. Clearly, the *edod* is also between 0 and 1. As the *dod*, the $edod = 0$ indicates that the system is static. Otherwise, $edod = 1$, the routing problem is totally dynamic.

By adding the time window, the effective degree of dynamism is denoted by $edod_{tw}$ and computed as follows:

$$edod_{tw} = \frac{1}{n_s + n_d} \sum_{i=1}^{n_s+n_d} \left(1 - \frac{r_i}{T}\right). \tag{11}$$

This measure takes into account the time windows' constraint of each customer, which is represented by $[a_i, b_i]$, where a_i and b_i are, respectively, the lower and upper time intervals.

In the same equation (Eq. 11), r_i is the reaction time, which is defined in [71] as *"the temporal distance between the time the request is received and the latest possible time at which the service of the requests could begin."* Hence, the reaction time of the ith real-time request is calculated as: $r_i = b_i - t_i$. Therefore, lower reaction time implies higher dynamic system and less flexibility for the planner to incorporate the new received request to the plan (as proven in [71]). Similar to the previous measure, the $edod_{tw}$ is between 0 and 1, without time window constraint ($a_i = t_i$ and $b_i = T$) $edod_{tw} = edod$.

4.3 Main Factors Leading to the Emergence of Dynamic VRP Variants

Real-life transportation scenarios often require vehicle routing that can adapt to changes over time. Changes in the environment, such as new demands and order cancelations, can lead to variations in the number of orders during the operation time in VRP. In such cases, routes have to be adapted to respond to the new environment conditions. As a result, DVRP differs from static VRP in several fundamental aspects [72, 104, 105].

The Concept of Time
The time means that the planner should know the position of the fleet of vehicles at any time. These to be able to inform the drivers about their next destination whenever new request appears.

Real-Time Information
In static VRP, all the inputs of the problem are known beforehand (before the vehicle leaves the depot) and have the same quality. However, in DVRP, customer requests are received in online fashion and the future information is unknown with certainty in real-life scenarios.

Problem Objective Function
In addition to minimizing the total travel cost, DVRP optimization can include other objectives, such as the number of customer requests and environmental aspect. Furthermore, in a highly dynamic system like the emergency service, the objective is to minimize the response time (i.e., the expected delay between the moment when the user's request arises and the moment when it is fulfilled).

Faster Computation Time
The computation time is fundamental in the DVRP unlike the static VRP, where the planner can spend a few hours to obtain a high-quality solution (even optimal one). In fact, the new appeared requests should be incorporated to the routing plan as fast as possible. In addition, each time the environment changes, the real-time algorithm has to provide the decision-maker with new routes in order to redirect the vehicle fleet as fast as possible.

Ability to Reject Requests
Customer rejection is more common in a dynamic routing than in a static VRP. This is due to the limited resources or hard time window constraints. Although exclusion of customers is generally not preferred, it can be a logical and obvious consequence of a hard time window constraint with a limited number of vehicles or a vehicle capacity constraint.

5 Main Variants of the DVRP

The nature of routing problems has evolved to become dynamic processes, and as the DVRP can model many real-world situations, more complex variants have been addressed over the years. These variants involve taking into account new constraints and objectives, which can make the problem more challenging to solve. Thus, a significant number of studies on DVRP and its variants have been published [102]. In what follows, we outline the main existing DVRP variants.

Capacitated DVRP (D-CVRP)
In this variant, the main objective is to minimize the total travel cost or distance. For instance, authors in [92] modeled DVRP as a series of VRPs. Then, they solved a static problem at each time slot that contains all the known customer demands that have not been served. In addition, a recent work studied in [140] suggested a demand coverage diversity adaptation approach to better deal with the environmental changes. The proposed method aims to guarantee the diversity of the population to better enhance the search space. Furthermore, authors in [1, 112] developed a GA to deal with the dynamic requests while respecting the vehicle capacity. The work of [95] studied the DVRP in the context of continuous search space. The particle swarm optimization (PSO) and differential evolution (DE) approaches are developed and tested. The authors in [64] have modeled the transportation of patients with a fleet of ambulances as a DVRP and solved the problem using a tabu search (TS) algorithm. Moreover, the problem of home delivery is modeled in [123] as a DVRP with linear programming (LP) model in order to minimize the travel cost. A queuing system with multiple stations is studied in [21], where a single class of customers is served. Each station possesses multiple servers and an exclusive queue. The manager of the system is responsible for determining the station to which each new customer should be directed.

Furthermore, several works recognized the D-CVRP as a dynamic multi-objective optimization problem (DMOOP). To address this issue, a bi-objective ant colony optimization (ACO) for the DVRP was introduced in [62]. Additionally, authors in [127] modeled the problem of courier mail services multi-objective DVRP. In their work, three objectives were considered: the minimization of customer waiting time, the minimization of the total travel time, and the maximization of the number of serviced customers. An LS is proposed to tackle this problem. The dial-a-ride problem (DARP) with transfers is modeled as DVRP in [101]. To solve the problem, large neighborhood search (LNS) and a MILP are proposed to minimize (i) the total travel distance, (ii) the number of vehicles used, and (iii) the number of unscheduled requests. A lexicographic multi-objective model is proposed in [144] to formulate the problem of dynamic ride-hailing sharing problem (DRHSP) with multiple vehicle types and user classes, in which three objectives have to be optimized: the profit, the number of requests, and the total driving distance. The authors designed an artificial bee colony (ABC) algorithm to solve the problem. The problem of EV is modeled as a DVRP in the work [137]. The authors developed an evolutionary multi-objective algorithm (M-EA) to optimize the charge cost and travel time.

DVRP with Time Windows (DVRPTW)

DVRPTW adds two additional constraints to DVRP to ensure that customers are served within their specific time intervals. Specifically, each customer is assigned a time window $[a_i, b_i]$, where a_i and b_i represent the earliest start and latest end of the service, respectively. The vehicle must arrive at each customer's location and begin service within this time window. The authors in [33] solved the problem using a variable neighborhood search (VNS). Other studies in this field belong to the work of [27] where an adaptive LNS (ALNS) is developed to serve a set of customers via a heterogeneous fleet of vehicles from one depot while minimizing the travel distance. Dealing with the same objective, an iterated LNS (ILNS) is developed in [54]. Moreover, a hybrid solver based on the multi-objective ACO to deal with real-world DVRPTW was developed in [142]. Besides, a GA is designed to solve DVRPTW in [11, 55]. A dynamic multi-objective optimization evolutionary algorithm (DMOEA) based on ensemble learning (EL) is developed in [136]. The goal of the addressed problem is to minimize the total customer waiting time and the travel distance. The work of [42] investigated the problem of dynamic Electric VRP. A GA is developed to minimize the total costs. Authors in [113] studied the problem of dynamic ride-sharing and taxi-sharing at the aim to maximize the number of served requests and to minimize the travel cost. A greedy randomized adaptive search procedure (GRASP) is designed to solve the problem. Similarly, [107] developed a heuristic approach based on a model predictive control (MPC). More recently, authors in [34] solved the problem of autonomous taxi request in hybrid mode (known and unknown demands) using CPLEX. The DARP with time windows (DARPTW) is investigated in [52]. The authors designed an adaptive insertion heuristic (AIH) to minimize the total cost. For the same problem, a benders decomposition (BD) is developed in [108]. Additionally, a bi-objective sustainable logistic model that minimizes the total travel distance and emissions is developed in [60] to deal with the real-time routing problem of cash-in-transit. The problem is solved by the nearest neighbor (NN) and the iterated local search (ILS) algorithm. Considering a fuzzy time windows besides the dynamic demands, [45] developed a GA in order to maximize customer preferences for service and to minimize: the number of used vehicles, the total travel distance, and vehicles waiting time. The work of [115] has taken into account time window constraints with split and delivery of heterogeneous fleets of vehicles (DVRPTWSD) to serve a set of customers. The ACO algorithm was developed to solve the studied problem. In the context of hazardous material transportation, authors in [98] modeled the problem as a bi-objective DVRPTW at the aim to minimize the total travel cost and the transportation risk. To solve the problem, the authors developed two multi-objective meta-heuristics: a bi-population GA and a hybrid approach combining the GA and the VNS (BIG-VNS and BI-GA). Due to the growing global concern about environmental issues, authors in [99] tackled the problem of toxic gas emissions caused by freight transportation. To this end, they developed a bi-objective model to reduce the CO_2 emissions and the transportation cost.

Pickup and Delivery DVRP

Another DVRP contribution relates to the pickup and delivery (D-PDVRP). For instance, authors in [36] developed an ACO as solution method. Moreover, a heuristic approach is proposed in [48, 141]. A MILP is proposed in [122] at the aim to maximize the total earliness to all customers, while also penalizing lateness to customers. A TS is suggested in [44].

In order to address the issue of D-PDVRP with time windows (D-PDVRPTW), a heuristic approach was proposed in [37, 63, 133]. To deal with the same problem, a VNS with multiple LS is proposed in [19]. Furthermore, the authors in [29] developed a GA to deal with the D-PDVRP. Then, a heuristic insertion is performed for the dynamic requests. The study presented in [39] introduced the concept of D-PDVRP with real-time control (DPDPRC) as a solution to the problem of courier service, at the aim of minimizing lateness at request locations and minimizing the operating costs of the used vehicles. To re-adapt routes, a TS algorithm guided by a multi-stage neighborhood operator selection scheme is developed.

The problem of autonomous vehicles is modeled as D-PDVRP in the work of [57] at the aim to minimize the travel distance and the traveler wait times. To solve the problem, the authors simulated six dynamic assignment strategies (DAS) for immediate requests. To deal with the same problem, authors in [74, 79] proposed and simulated a general framework. Furthermore, authors in [138] modeled the problem of disaster relief as a D-PDVRP. A TS algorithm is designed to minimize the total travel time and the number of used vehicles.

The problem of the dynamic taxi shared is studied in the work of [56]. The objective of the studied problem is to maximize the total profit of the system. To do, the authors solved the problem to optimality using a branch and price (B&P) algorithm and a Lagrangian relaxation (LR).

Stochastic DVRP

The stochastic DVRP (DSVRP) is a variant of the VRP that incorporates both dynamic and stochastic elements. In the DSVRP, the objective is to optimize the routing and scheduling of vehicles to serve customers with time-varying and uncertain demands. The dynamic aspect of the problem refers to the changing nature of customer demands over time. This means that the demand for goods or services may vary throughout the planning horizon, requiring adaptive routing decisions to meet customer requirements effectively. The stochastic aspect of the problem relates to the uncertainty associated with the demand. Thus, the demand for goods or services is not known with certainty but follows a probabilistic distribution. This uncertainty can be due to various factors such as fluctuating customer demand, unpredictable order cancelations or additions, or random variations in customer requirements. It is worth noting that the specific formulation and characteristics of the DSVRP may vary depending on the problem context, research objectives, and modeling assumptions employed in different studies [110, 132]. For instance, the study presented in [117] proposed a novel approach to emergency evacuation. The problem is formulated as a probabilistic multi-destination DVRPTW (PMD-DVRPTW) at the aim to maximize the number of evacuees transported from assembly points to designated shelters using the safest routes. To solve this problem, the authors developed a greedy algorithm. In [84], the authors aimed to address the dynamic bus-routing problem with stochastic passenger demand. To minimize the cost of vehicle travel time and penalty for rejecting requests, they proposed a two-stage stochastic programming model (T-SPM). To handle the dynamic changes, they adopted a rolling horizon technique. Additionally, they developed a vector-similarity-based clustering ALNS (VSC-ALNS) to solve the problem. In the context of dynamic and stochastic patient transportation problem, authors in [111] modeled the problem as DSVRP. The authors developed a VNS to minimize the overall lateness and the total travel time for the vehicles. The work of [47] investigated the problem of stochastic D-PDVRP (SD-PDVRP). An anticipatory heuristic that anticipates future demands through a Monte Carlo sampling procedure is designed to maximize the overall customer service level. Authors in [82] considered the stochastic environment in the DARP (S-DARP) to minimize the total travel cost. The problem is solved using an ALNS. The study presented in [148] addressed the DVRP with stochastic customer requests (DVRPSR) at the aim to maximize the number of served demands using MDP and approximate dynamic programming (ADP). To deal with the same problem, a heuristic approach based on multiple-knapsack approximation (M-KA) is developed in [148]. Similarly, [41] proposed a heuristic approach based on a multi-agent system integrated with trajectory data mining techniques (MAS-DM). The same method is developed in [131] for the dynamic multi-period VRP with stochastic service requests (DS-PVRP) in order to maximize the number of accepted requests. Considering the uncertainty in the service time, authors in [70] developed a fuzzy ACO to maximize the number of served customers and minimize the average of customer's waiting time.

The work presented in [12] focused on the dynamic stochastic electric VRP (DS-EVRP), with the goal of minimizing expected energy consumption. The authors proposed a safe reinforcement learning (SRL) combined with TS to solve the problem. The DVRPSR is also investigated and solved by an offline approximate value iteration (OAVI) in [147] to minimize the total expected cost for all vehicles to serve all customers. In the context of bike-sharing systems, [18] modeled the problem as a stochastic-dynamic inventory routing problem (SDIRP), with the objective of avoiding unsatisfied demand by dynamically relocating bikes during the day. They proposed an ADP approach to solve the problem. The study presented in [73] focused on the schedules of the zonal-based flexible bus service (ZBFBS), considering stochastic demand and location, time-dependent travel time, and passenger time window constraints. The problem is solved using a gradient-based solution (GBS) approach. Finally, the work of [87] modeled the problem of target surveillance as an SDVRP, and they proposed an MDP as a solution method. A recent survey for the DSVRP can be found in [120].

Multi-Depot DVRP

Based on the conventional multi-depot VRP, the dynamic version (D-MDVRP) arises in the case when not all orders are known before the departure of vehicles from the set of depots. In this context, authors in [116] developed a TS to determine a set of optimal routes for a fleet of vehicles originating from several depots taken into account the new occurred requests. Besides, [89] developed a heuristic approach in order to minimize the travel distance of the D-MDVRP. The time-dependent multi-depot green VRPTW is studied in [38]. The authors designed a hybrid GA with VNS to minimize the total costs.

Time-Dependent DVRP

In this variant, the travel time between two customers is considered variable through the time. This variation is due to weather conditions, traffic jam. For instance, [38, 51] investigated the DVRP with time-dependent travel times. To solve the studied problem, a GA is implemented as a solution method. In addition, a time-dependent travel time is addressed in the work of [25], besides the real-time information that occurs during the travel horizon.

6 Dynamic Multi-Objective Routing Problem

Real-world applications are dynamic in nature due to the change of the problem features over time. For instance, a new request appears, change of routes due to weather conditions or traffic jam. Therefore, evolution of the problem can affect the objective function and/or the system constraints. The changes in the system constraints are related to the input of the problem (resources), accordingly the feasibility of solutions is modified [26, 139]. Furthermore, real-life applications are not only dynamic in nature, but also require the simultaneous optimization of multiple conflicting objectives [9, 78]. Thus, decision-makers are not only focused on achieving an economic gain but also on a solution that can adapt the routing plan when a change occurs such as vehicle breakdowns, varying customer numbers, or route closures. Additionally, in DMOOPs minimizing the economic factors alone may not be sufficient. It is crucial to track the set of Pareto optimal solutions after environmental changes, which is a difficult task due to conflicts between multiple objectives and changing objective functions over time. Ensuring the best compromise between multiple objectives in a dynamic environment is thus a challenging and crucial task.

Formally speaking, DMOOP can be defined as follows: Given an optimization algorithm A to solve a dynamic problem f_t during an optimization period t where $t \in T$, if during, T the underlying fitness landscape that A uses to represent f_t

evolves and, therefore, A has to react to this update in order to provide new optimal solutions [9, 93].

In what follows, we first introduce few concepts related to DMOOPs. Without loss of generality, a minimization problem is considered. Formally, a DMOOP can be defined as [61]

$$\begin{cases} Min \quad F(x,t) = (f_1(x,t), f_2(x,t), \ldots, f_I(x,t)) \\ s.c \\ g_l(x,t) \geq 0 \qquad \forall l \in \{1, \ldots, L\} \\ h_m(x,t) = 0 \qquad \forall m \in \{1, \ldots, M\} \end{cases}, \tag{12}$$

where F represents the set of $I \geq 2$ objectives to be minimized, and g and h denote the inequality and the equality constraints over time t, respectively.

Definition 1 (Dynamic Pareto Dominance) Given two vectors x_1 and x_2, we said that $x_1 \prec x_2$, if and only if we have $f(x_1, t) \leq f(x_2, t)$, $\forall i \in \{1 \ldots N\}$, and $\exists i \in \{1 \ldots N\}$, such that $f(x_1, t) < f(x_2, t)$.

Definition 2 (Dynamic Pareto Optimal Solutions (DPOS)) A set of solutions is called a DPOS at time t represented by $DPOS(t)*$, if and only if the following conditions are validated:

$$DPOS(t) = \{(x \in D | \neg\exists x' \in D, f(x', t) \prec f(x, t)\}.$$

Definition 3 (Dynamic Pareto Optimal Front) The $POF(t)$ for a given MOOP satisfies the following condition:

$$POF(t) = \{f(x, t), x \in POS\}.$$

7 Solution Methods for the DVRP

A wide variety of solution methods have been developed and used to solve the static VRP. Also, the spectrum of possible solution methods for DVRPs is equally wide [88, 102, 105, 110]. Considering the fact that fast solution times are crucial, the major part of the existing approaches are approximate methods. Indeed, the DVRP is an NP-hard problem [102], and exact methods only provide an optimal solution (with small problem instances) for the current state, without necessarily guaranteeing that the solution will remain optimal as new information becomes available [46].

Furthermore, the existing solution methods are designed based on two important dimensions [105]. The first is the *evolution of information*, which means that in the static case, all problem inputs are revealed to the planner before the routes are executed (before the vehicles leave the depot). However, the information of the dynamic environment is susceptible to change during the working day (e.g., a new demand appears) and must be integrated into the planning in order to satisfy all the orders. The second dimension is related to the *information quality* in which the input information could be (i) deterministic (known with certainty) and (ii) stochastic (uncertainty). To this end, we can broadly classify the DVRP problem into two main categories, as follows:

1. *Dynamic and deterministic*: In this class, all or part of inputs are revealed to the planner over time with certainty. In fact, deterministic inputs denote that when a new request appears, all its information (location, demand, service time) is known with certainty such as the work of [48, 97–99, 122, 141].
2. *Dynamic and stochastic*: The dynamic request is known with some probabilistic information. In other words, the exact value of the request that dynamically evolves became known only when the vehicle arrives at the customer. For further details about the dynamic and stochastic VRP, the reader is referred to [110, 114].

The solution approaches designed for the DVRP can be broadly classified into two main categories: ones involving periodic re-optimization and ones involving continuous re-optimization [102]. In the rest of this section, we outline the main features of each category, and we provide an overview of the main existing works, besides some other existing strategies dealing with the DVRPs.

7.1 Periodic Re-optimization

The periodic re-optimization approaches consist in dividing the working day into time slots. The idea behind dividing the working day into a set of time slots is introduced in [66] at the aim to specify the time allowed for each static problem. In each time slot, the problem corresponds to static VRP. Indeed, a $t = 0$, where $t \in T$ (T represents the length of the working day), corresponds to the initial static problem that contains all the known customers in advance. The main benefit of periodic re-optimization is that it is based on algorithms that have been developed for the static VRP. To the best of our knowledge, the first periodic re-optimization algorithm was introduced in [103]. After that, several solution methods have been proposed, such as the GA that has been successfully applied to solve the periodic re-optimization DVRPs. For instance, this meta-heuristic is used in the work of [1] to minimize the total travel distance. Besides, an adaptive version of the GA is developed in [10]. Additionally, a self-adaptive evolutionary algorithm (SA-EA) is proposed in [112]. Moreover, a MA is used in the work of [86]. Authors in [98] developed two multi-objective meta-heuristics: a bi-population GA and a hybrid approach combining the GA and the VNS (BIG-VNS and BI-GA). Furthermore, among the bio-inspired algorithms that have been proposed to solve the DVRPs is the multi-swarm optimizer (MSO) [94], also, the ACO in [36, 62, 70, 92, 140, 142, 144]. Furthermore, many trajectory-based meta-heuristics are developed. For instance, the ALNS is designed in [8, 27, 82], the ILNS in [54], the VNS in [65], NN and ILS in [60], and the TS in [39]. Further, a dynamic CG is developed in [24] for the DVRPTW to minimize the total travel distance. Also, a heuristic approach based on a multi-agent system integrated with trajectory data mining techniques (H-MADM) is proposed in [41]. Additionally, a heuristic approach based on M-KA is designed in [148]. Finally, a GRASP is developed in [113].

7.2 *Continuous Re-optimization*

Continuous re-optimization is performed during the working day. In every environmental changes, the new appeared orders are aggregated with the none yet serviced requests to update the current routing plan. The first continuous re-optimization is introduced in [43]. In their work, authors adapted the TS for DVRPTW by the use of an adaptive memory in order to check the feasibility to integrate the new occurred orders. After that, several works have been studied, and many solution methods have been developed. By studying the existing works, it is noticeable that EAs and their variants are the most investigated solution methods. For instance, the GA is used in the work of [10, 11, 45, 51, 55] at the aim to minimize the total travel distance. A multi-objective vector-evaluated EA augmented with an exploitation phase and hyper-mutation (VEEA) is proposed in [99]. Moreover, two meta-heuristics: PSO and DE are developed in [95]. Additionally, the PSO is developed in [65] to solve continuously the DVRP while minimizing the total travel distance. Further, several solution-based meta-heuristics are adapted to continuously re-optimize routes in order to incorporate the new appeared requests such as a VNS in [33] to for the DVRPTW to minimize the total travel distance, and they applied this approach in a real case study. Additionally, a parallel TS is suggested in [44] for the pickup and delivery requests that occur in online fashion. The same approach is also addressed in the work of [39, 58]. Finally, an heuristic approach is proposed in [80].

7.3 *Other Strategies*

In addition to the methods previously discussed, there are several other strategies to handle online requests. These methods are divided into *simple policies* or an *insertion technique* [104]. In what follows, we elucidate the main existing solutions strategies.

Insertion Heuristic
In this strategy, the occurred requests have to be inserted mainly in the best position (also known cheapest insertion) in the already created route. Thus, all possible positions are examined, and then the customer is inserted in the adequate order. It should be noted that, in some cases, dynamic requests are inserted into the routing

plan and the visit orders are then refined by other solution methods (e.g., TS). The insertion strategy is widely used since it was introduced for the DVRP in [59]. For instance, authors in [73, 84, 103] developed an insertion heuristic that is applied in a rolling horizon. Moreover, [90] proposed a double-horizon-based heuristics to deal with the D-PDVRP. The same method is designed in the context of DARP in [22] to minimize the total travel time while serving a set of known and unknown customers. In [19, 25, 87] developed this technique to deal with the new appeared customers. In the context of patient transport, the authors in [13] have applied this approach to incorporate the new demands that arise, and then the routing plan (solution) is improved by the TS. Moreover, an insertion heuristic is designed in [29]. An adaptive insertion heuristic is developed in [52]. The same approach is used in [63]. Rollout heuristics are implemented in [12] for the DS-EVRP. Authors in [48] proposed a scalable anticipatory policy (SAP) for the D-PDVRP. Considering the stochastic customer request for the D-PDVRP, an anticipatory heuristic based on Monte Carlo sampling procedure (AH-MCP) is proposed in [47]. In the work of [138], the authors designed an anticipatory heuristic for the travel time for the D-PDVRP. Furthermore, a proactive heuristic is proposed in [122] in order to deal with dynamic requests for the DARP. To deal with the problem of autonomous vehicle, a dispatching-based heuristic strategy is proposed in [34], while in [57] six DAS for immediate demands were developed. In the work of [52], an adaptive insertion heuristic is proposed for the D-VRPTW. A heuristic insertion based on an MPC is proposed in [107] for the dynamic taxi-sharing.

First Come First Served

In this technique, requests are treated in the order in which they are received [14]. In the work of [51], the authors used the first come first served to deal with the dynamic requests while considering the travel time as time-dependent. In addition, the same policy is addressed in [141] to deal with the online pickup and delivery requests.

Nearest-Neighbor Policy

This strategy assigns the request that has appeared to the vehicle closest to its position. It should also be noted that if a request appears, while the vehicle is traveling between customers i and j, then the scheduling plan is updated after serving customer j [14]. To deal with the dynamic traveling repairman problem (DTRP), the authors in [71] used the nearest-neighbor policy (NNP). Also, authors established a comparison of the considered strategy against other ones to minimize the routing cost. A refined version of the nearest neighbor is proposed in [118] to deal with the DTRP. The primary feature of their approach involves rerouting or reassigning vehicles that may already be on route.

Traveling Salesman Policy

Requests are grouped into sets of a given size. Once a set of requests has been collected, a TSP is solved. Requests are then processed according to the optimal solution of the TSP [14].

Tables 3 and 4 provide an elaborate taxonomy of the literature dealing with DVRPs in the single and multi-objective optimization, respectively.

Table 3 Review of existing literature on DVRP for single-objective optimization

Optimization goal	Variant	Solution approach	Reference
Distance/Cost/Time	D-CVRP	AGA	[10]
		SA-EA	[112]
		MA	[86]
		PSO and VNS	[65]
		PSO and DE	[94]
		ACO	[92, 140, 142]
		Insertion heuristic	[59]
		ALNS	[8]
		ILNS	[54]
		GA	[1, 42]
		LP	[123]
		TS	[13, 64]
	D-VRPTW	VNS	[33]
		ALNS	[27]
		ILNS	[11, 55]
		AIH	[52]
		BD	[108]
		CPLEX	[34]
		GA	[54]
		CG	[24]
		ACO	[36]
	D-PDVRP	TS	[44]
		B&P and LR	[56]
		Simulation	[74, 79]
		Heuristic	[48, 141]
		GA and heuristic	[29]
	DVRPSR	MDP and ADP	[148]
		M-KA	[148]
		MAS-DM	[41, 131]
		OAVI	[147]
		MDP	[87]
	D-MDVRP	MDP	[116]
		Heuristic	[89]
	D-PDVRPTW	Heuristic	[37, 63, 133]
		VNS	[19]
	DTRP	NNP	[71]
	TD-VRP	GA	[51]
	ZBFBS	GBS	[73]
	SDIRP	ADP	[18]
	DVRPTWSD	ACO	[115]
	D-MDVRP	TS	[116]
	S-DARP	ALNS	[82]
Number of evacuees	PMD-DVRPTW	Greedy algorithm	[117]
Energy consumption	DS-EVRP	SRL and TS	[12]

Table 4 Summary of the existing literature on DVRP in multi-objective optimization

Variant	Optimization goal	Solution approach	Reference
D-CVRP	– Cost	M-EA	[137]
	– Travel time		
	– Customer waiting time	LS	[127]
	– Total travel time		
	– Number of serviced customers		
DARP	– Number of vehicles	MILP and LNS	[101]
	– Distance		
	– Number of unscheduled requests		
DRHSP	– Profit	ABC	[144]
	– Distance		
	– Number of requests		
	– Cost	GRASP	[113]
	– Number of requests		
DVRPTW	– Customer waiting time	EL	[136]
	– Total travel distance		
	– Distance	VEEA	[99]
	– CO_2 emissions		
	– Cost	BIG-VNS and Bi-GA	[98]
	– Transportation risk		
D-PDVRP	– Total distance	DAS	[57]
	– Wait times		
	– Total earliness	MILP	[122]
	– Penalizing lateness to customers		
	– Travel time	TS	[138]
	– Number of vehicles		
DPDPRC	– Lateness at request	TS	[39]
	– Cost		
DS-PVRP	– Number of served customers	Fuzzy ACO	[70]
	– Average of customer's waiting time		
DBRP-SD	– Cost	T-SPM and VSC-ALNS	[84]
	– Penalty for rejecting requests		
DVRPSR	– Overall lateness	VNS	[111]
	– Travel time		
Fuzzy DVRPTW	– Customers preferences for service	GA	[45]
	– Number of vehicles		
	– Distance		

8 Performance Measures

To better evaluate the behavior of an algorithm, it is essential to quantify its performance and the quality of the obtained solutions. In the literature, there is a considerable number of performance measures dealing with static and dynamic problem, single and MOOP. In the case of single-objective optimization, it is usually sufficient to assess the final solution [9]. However, in the dynamic optimization, these measures are required to evaluate the effectiveness of the solution method, regarding the environmental changes, to allow a meaningful comparison and to draw balanced conclusions. Among the existing metrics, we can cite:

Value of information (Val_Inf) provides the gap between a static and a dynamic solution of such algorithm. Accordingly, it quantifies the impact of the dynamism on the performance of an algorithm. In other words, it calculates the effectiveness of an algorithm to solve a dynamic problem [90] and is calculated as follows:

$$Val\_Inf = \frac{Z_{Alg(I_d)} - Z_{Alg(I_s)}}{Z_{Alg(I_s)}} \times 100, \tag{13}$$

where $Z_{Alg(I_d)}$ is the obtained solution of such approach Alg in a dynamic instance I_d and $Z_{Alg(I_s)}$ represents the value of the objective function achieved by an algorithm Alg for the static instance I_s.

Average relative percentage deviations (ARPD) measures the average deviation of an algorithm compared to the best obtained solution based on the same instance. The ARPD is given as follows:

$$ARPD(a) = \frac{ObtSol(a) - BestSol*}{BestSol*}, \tag{14}$$

where $BestSol*$ denotes the best known solution and $ObtSol(a)$ is the fitness value obtained by the algorithm a in the same instance.

Diversification Matrix (DM) is a performance measure that evaluates the diversity of solutions in the POF. It calculates the Euclidean distance between each pair of solutions in the POF and then computes the average of these distances. A higher value of DM implies that the algorithm has a better capability in terms of diversification. The equation for calculating DM is given in Eq. 15 [7]:

$$DM = \sqrt{(maxf_{1i} - minf_{1i})^2 + (maxf_{2i} - minf_{2i})^2}. \tag{15}$$

In this equation, $maxf1i$ and $minf1i$ indicate the maximum and minimum values of the ith objective function achieved by non-dominated solution, respectively.

Mean Ideal Distance Measure (MID) represents the convergence rate of POF to a given point, typically the ideal point in multi-objective optimization [50]. It assesses the closeness between solutions in the POF and the ideal point. Without loss of generality, in the case of minimization, the ideal point has the coordinate (0, 0). The MID is calculated as follows:

$$MID = \frac{\sum_{i=1}^{n} \sqrt{f_{1i}^2 + f_{2i}^2}}{n}. \tag{16}$$

In Eq. 16, n denotes the number of non-dominated solutions and f_{1i} and f_{2i} represent the first and second objective values of the ith non-dominated solution, respectively.

Pareto Dominance Indicator (PDI) measures the quality of the set of non-dominated solutions found by an algorithm compared to the set of non-dominated solutions found by all algorithms. A higher value of PDI indicates that the algorithm set has found a larger number of solutions that are non-dominated by all algorithms, which is preferred over a lower value [49]. The formula for PDI, in Eq. 17, is as follows: The PDI metric is formulated in Eq. 17 and as it is clear that higher values are preferred to lower ones

$$PDI(X_1, X_2, \ldots, X_m) = \frac{|X_i \bigcap Y|}{|Y|}, \tag{17}$$

where $Y = \{y_i \,|\, \forall y_i, \neg\exists x_j \in (X_1 \cup X_2 \cup \ldots \cup X_m) < y_i\}$, and $x_j < y_i$ implies that x_j dominates y_i. X_i is the set of solutions under evaluation.

Ratio of Non-dominated Individuals (RNI) metric is used to assess the proportion of non-dominated individuals in the population X. The larger the value of RNI, the better the algorithm. A value of RNI equal to 1 indicates that all the solutions in the population are non-dominated, while a value of 0 indicates that all the solutions are dominated. The RNI is given in the following formula [126]:

$$RNI = \frac{|\bar{X}|}{P}. \tag{18}$$

Here, $\bar{X}$ is the set of non-dominated solutions, and P is the total size of the population. The RNI value falls within the range [0, 1]. An RNI value of 1 indicates that all solutions in X are non-dominated, while an RNI value of 0 implies that all individuals in X are dominated.

Normalized Score (NS) metric is used to compare algorithms efficiency through different problem instances and/or many change period (for dynamic problem) [93]. The NS calculates the performance of the ith algorithm by normalizing the results given by each algorithm to the range (0, 1). Through this metric, the best performing algorithm will get a higher overall score. Otherwise, it will get less overall score. The NS formula of the ith algorithm is presented as follows, in Eq. 19:

$$NS(A,i) = \frac{1}{N}\sum_{i=1}^{N} \frac{|\ (Alg(A,i) - MIN(i))\ |}{|\ (MAX(i) - MIN(i))\ |}. \tag{19}$$

Non-dominated Solutions (NDS) performance measure is calculated by counting the number of Pareto Optimal Fronts (POFs) obtained by each algorithm [4]. A higher value of NDS indicates a better performance of the algorithm.

9 Benchmarks for the DVRP

To date, there has been a remarkable absence of comprehensive benchmark conducted for the DVRP. Accordingly, existing research in this field has relied on the use of static benchmarks and their transformation into dynamic data sets to address the problem at hand. The main used static benchmarks are Solomon's benchmarks [121] and Kilby et al. [66]. By simulating dynamic aspects of the problem, these benchmarks provide researchers with a more realistic environment to evaluate and compare different algorithms and strategies. In what follows, we detail these two data sets.

The benchmarks of Solomon.
One of the primary sources of static benchmarks that have been adapted for the DVRP is Solomon's benchmarks [121]. Initially designed for the static VRPTW, these benchmarks consist of various problem instances with different features. It includes a total of 56 instances containing 100 customers and one central depot. Also, each benchmark instance includes: fleet size, vehicle capacity, traveling time of vehicles, and spatial and temporal distribution of customers to be served. These instances are classified into three categories: R-type (uniformly distributed customers), C-type (clustered customers), and RC-type (a mix of R- and C-type). Problems in categories R, C, and RC are further sub-classified as $R1$, $C1$, $RC1$, $R2$, $C2$, and $RC2$. Problem sets $R1$, $C1$ and $RC1$ have a short scheduling horizon and allow only few customers per route, while problems in sets $R2$, $C2$ and $RC2$ have a long scheduling horizon, permitting many customers to be served by the same vehicle. These problems differ with respect to the width of the time windows sizes.
To incorporate dynamic elements into Solomon's instances, researchers often applied the dod (Eq. 9).

The benchmarks of Kilby et al. [66].
Another significant source of static benchmarks used in DVRP is proposed by Kilby et al. [66]. These data sets are derived from the well-known VRP benchmarks of [125] (12 data sets), [30] (7 data sets), and [40] (2 data sets). These problems ranged from 50 to 199 customers. To turn these static data sets into dynamic ones, the authors in [66] incorporated the concept of the length of the working day. Also, the appearance time of each request is added. This

represents the time when the demand becomes available during the working day. In addition, a duration for each task, which means the required time to serve a request. Furthermore, the number of available vehicles was fixed as a large number to cover all requests. In addition, the speed of vehicles is assumed to be equal to 1.0 (unit of distance/unit of time). Let us notice that these of benchmarks are without time windows. Additionally, the number of customers is indicated in the name of each instance (tai50a means a static VRP problem from Taillard's benchmark, containing 50 customers).

10 Real-World Applications of the DVRP

The VRP, particularly the DVRP, enables the modeling of various applications in the transportation and distribution domain. Below, we outline the significant applications where dynamic routing has been successfully implemented.

Courier Services
In practice, courier services are provided by several companies (e.g., FedEx, Federal Express). The picked up mails are loaded from several locations and have to be delivered to another destinations. The dynamic aspect of the problem can be seen in the pickup request since online demands occur for service; therefore, the driver or the dispatcher has information on when and where the pickups are going to take place. However, the delivery task is mainly considered as a static routing problem since all destinations are known by the driver [77, 127]. For instance, authors in [39] studied the problem of courier service.

Dial-a-Ride Systems
This routing problem can be seen as one of the most applications of the pickup and delivery VRP for urban areas. An example of the dynamic variant of the DARP is the transportation of elderly or handicapped people [85]. The DARP in the case of arising in Austrian rural is studied in [5]. Moreover, the problem of home delivery is modeled in [123]. Furthermore, the problem of automated taxis in [76], Taxi and ride-sharing in [113]. Further, several algorithms have been developed for this problem, such as: [22, 52, 59, 80, 82, 85, 101, 108, 111, 124]. For more details about the DARP, readers are invited to see [53, 72].

Electric Vehicle
The use of the DVRP presents a robust method for enhancing the routing and overall performance of EV fleets, resulting in more efficient and eco-friendly operations. In recent literature, various studies have focused on the potential of EVs to reduce emissions. Furthermore, EVs have the advantage of saving on transportation costs due to the cheaper cost of electricity compared to gasoline or diesel fuel, as well as requiring less maintenance than conventional vehicles. Among the recent works in the literature, we can cite [12, 28, 42, 75, 128, 135, 137, 149].

Emergency Services
The services that fall under the category of emergency services, such as fire, police, and ambulance services [143], are highly dynamic in nature ($dod = 1$) since all requests are received in real time. Typically, routes are not re-optimized, as requests are usually served before a new one appears. The approach in this type of service is to assign the incoming request to the best-suited (nearest) vehicle in order to serve it as quickly as possible. Thus, emergency services usually rely on location analysis to determine the best location for the emergency vehicle [67].

For instance, notable research studies have concentrated on addressing the challenge of patient transportation using a fleet of ambulances [64, 111]. Moreover, an innovative method for emergency evacuation was introduced in [117]. Additionally, the issue of disaster relief was modeled as a D-PDVRP in [138].

Public Transport Services—Taxi Cab Services
Real-life routing problems are prevalent in several areas, including taxi cabs, bus routing, and bike-sharing systems. In the taxi cab, the policies employed to assign customers to taxis can differ across companies. Additionally, taxis have the capability to serve multiple customers simultaneously [76]. The problem of dynamic ride-sharing and taxi-sharing is studied in [56, 113]. The dynamic bus routing with stochastic passenger demand is addressed in [84]. The study is presented in [73] focused on the scheduling of zonal-based flexible bus services. For bike-sharing systems, the problem is formulated as a SDIRP in [18], with the objective of avoiding unsatisfied demand by dynamically relocating bikes during the day.

Hazardous Materials Transportation
The VRP has emerged as a critical tool for transporting hazardous materials safely, given the risks associated with it for both human health and environment [129]. In recent years, this problem has gained significant attention due to its potential for fatal accidents during transportation. To address this, the authors in [98] conducted the first study on the dynamic variant of the hazardous material transportation, which considers changes in the environment that may affect preplanned routes. In their study, the authors put forward a bi-objective optimization model with the goal of minimizing total transportation costs while simultaneously mitigating travel risk. They also introduced a novel risk formula to quantify the level of risk associated with the transportation routes.

11 Trends and Future Directions of Dynamic Vehicle Routing Problem

There are several potential future directions in the field of the DVRP. Some of the key challenges and opportunities in this field include:

- Sustainable development has become a pressing concern in the field of transportation, prompting increasing attention to the integration of sustainability criteria into the DVRP. The incorporation of sustainability considerations into DVRP has led to the development of new routing algorithms that aim to minimize environmental impacts, such as fuel consumption and emissions, while maintaining the quality of service. One trend in this area is the development of multi-objective optimization models that consider sustainability as a primary objective. This approach involves the simultaneous optimization of multiple objectives, such as cost, travel time, and emissions, while balancing trade-offs between them. Among the limitation is the lack of real-world implementation of sustainability-focused DVRP models due to the challenges of collecting and managing data on environmental impacts and operational costs. Another limitation is the trade-offs between sustainability criteria and other objectives, such as customer satisfaction, that may impact the quality of service. The only work that considers simultaneously the minimization of the travel cost and the CO_2 emissions is studied in [99].

- The integration of EV into DVRP has become a popular research topic due to the growing demand for sustainable transportation solutions. The challenges of incorporating EV into DVRP include limited range, charging infrastructure availability, and battery degradation. Trends and future directions of DVRP with EV include the use of machine learning and artificial intelligence techniques to optimize routing decisions, the integration of Internet of Things technology to monitor EV performance in real time, and the use of vehicle-to-grid (V2G) technology to enable EV to feed energy back into the grid.
- Anticipating new requests with machine learning is an important problem in various fields and can be tackled with different techniques such as time-series forecasting, clustering, collaborative filtering, association rules, and deep learning. By using historical data, machine learning can be used to predict the number and type of requests that are likely to be made in the future, and appropriate actions can be taken to meet those requests in a timely and efficient manner.
- From a computational perspective: incorporating parallel computing and high-performance computing techniques. The future of computational perspectives in DVRP will focus on the development of algorithms that handle real-time request and dynamic routing decisions while incorporating uncertainty and risk management. These advancements can improve efficiency and effectiveness in transportation and logistics while offering new opportunities for developing effective solutions for DVRP to solve complex data sets while meeting multiple objectives.
- The DVRP has a lack of benchmark data sets that represent the dynamic nature of the problem. Currently, most existing data sets used to test DVRP algorithms are adaptations of static benchmarks, which limits the evaluation of the algorithms' performance. The future of DVRP researches will involve the development of benchmark data sets that represent real-world dynamic scenarios to improve the evaluation and comparison of algorithms. These data sets should incorporate online demand information, dynamic traffic conditions, and other uncertainties that affect the routing decisions. Additionally, the development of benchmark

data sets will require the cooperation of stakeholders to collect and share data, which can facilitate the development of more effective DVRP algorithms that can handle the complexities of real-world dynamic scenarios. For instance, the use of smart bin in the context of waste transportation, the IoT to deal with the real-time requests.

12 Conclusion

The DVRP has become a vital area of research in the fields of logistics and transportation, receiving significant attention over the past few decades. The aim of this chapter is to present a comprehensive survey of the literature devoted to the DVRP. To begin, we introduced the VRP and its main variants. Next, we outlined the essential features of the DVRP and highlighted the most significant differences between static and dynamic routing problems. Additionally, we provided an overview of the existing variants and solution methods for the DVRP.

Despite the potential benefits of the DVRP, it has several limitations that hinder its practical implementation. One of the most significant challenges is the high computational complexity of DVRP algorithms, which can make real-time implementation difficult. Furthermore, the uncertainty is related to several factors such as demand and traffic conditions. The limited availability of dynamic data sets also presents a challenge, making it difficult to evaluate and compare DVRP algorithms in real-world scenarios. To address these limitations, future research should focus on developing more efficient algorithms capable of handling the complexities of real-world scenarios, incorporating sustainable solutions such as EV in multi-objective framework. Also, improving the availability of dynamic data sets to enable the evaluation and comparison of DVRP algorithms in realistic scenarios.

References

1. A.M.F. Abdallah, D.L. Essam, R.A. Sarker, On solving periodic re-optimization dynamic vehicle routing problems. Appl. Soft Comput. **55**, 1–12 (2017)
2. S. Aday, M.S. Aday, Impact of covid-19 on the food supply chain. Food Quality Safety **4**(4), 167–180 (2020)
3. M. Amous, S. Toumi, B. Jarboui, M. Eddaly, A variable neighborhood search algorithm for the capacitated vehicle routing problem. Electr. Notes Discrete Math. **58**, 231–238 (2017). 4th International Conference on Variable Neighborhood Search
4. M. Arjmand, A.A. Najafi, Solving a multi-mode bi-objective resource investment problem using meta-heuristic algorithms. Adv. Comput. Techniq. Electromag. **1**, 41–58 (2015)
5. P. Armbrust, P. Hungerländer, K. Maier, V. Pachatz, Case study of dial-a-ride problems arising in Austrian rural regions. Transp. Res. Procedia **62**, 197–204 (2022)
6. F. Arnold, M. Gendreau, K. Sörensen, Efficiently solving very large-scale routing problems. Comput. Oper. Res. **107**, 32–42 (2019)

7. H. Asefi, F. Jolai, M. Rabiee, M.E. Tayebi Araghi, A hybrid NSGA-II and VNS for solving a bi-objective no-wait flexible flowshop scheduling problem. Int. J. Adv. Manuf. Technol. **75**(5), 1017–1033 (2014)
8. N. Azi, M. Gendreau, J.Y. Potvin, A dynamic vehicle routing problem with multiple delivery routes. Ann. Oper. Res. **199**(1), 103–112 (2012)
9. R. Azzouz, S. Bechikh, L. Ben Said, *Dynamic Multi-objective Optimization Using Evolutionary Algorithms: A Survey* (Springer International Publishing, Cham, 2017), pp. 31–70
10. M. Barkaoui, M. Gendreau, An adaptive evolutionary approach for real-time vehicle routing and dispatching. Comput. Oper. Res. **40**(7), 1766–1776 (2013)
11. M. Barkaoui, J. Berger, A. Boukhtouta, Customer satisfaction in dynamic vehicle routing problem with time windows. Appl. Soft Comput. **35**, 423–432 (2015)
12. R. Basso, B. Kulcsár, I. Sanchez-Diaz, X. Qu, Dynamic stochastic electric vehicle routing with safe reinforcement learning. Transp. Res. Part E Logist. Transp. Rev. **157**, 102496 (2022)
13. A. Beaudry, G. Laporte, T. Melo, S. Nickel, Dynamic transportation of patients in hospitals. OR Spectrum **32**(1), 77–107 (2010)
14. D.J. Bertsimas, G. Van Ryzin, A stochastic and dynamic vehicle routing problem in the Euclidean plane. Oper. Res. **39**(4), 601–615 (1991)
15. S. Braaten, O. Gjønnes, L.M. Hvattum, G. Tirado, Heuristics for the robust vehicle routing problem with time windows. Expert Syst. Appl. **77**, 136–147 (2017)
16. K. Braekers, K. Ramaekers, I. Van Nieuwenhuyse, The vehicle routing problem: state of the art classification and review. Comput. Ind. Eng. **99**, 300–313 (2016)
17. J. Brandao, Iterated local search algorithm with ejection chains for the open vehicle routing problem with time windows. Comput. Ind. Eng. **120**, 146–159 (2018)
18. J. Brinkmann, M.W. Ulmer, D.C. Mattfeld, Dynamic lookahead policies for stochastic-dynamic inventory routing in bike sharing systems. Comput. Oper. Res. **106**, 260–279 (2019)
19. J. Cai, Q. Zhu, Q. Lin, Variable neighborhood search for a new practical dynamic pickup and delivery problem. Swarm Evolut. Comput. **75**, 101182 (2022)
20. A.M. Campbell, J.H. Wilson, Forty years of periodic vehicle routing. Networks **63**(1), 2–15 (2014)
21. P. Cao, Z. Zhong, J. Huang, Dynamic routing in a distributed parallel many-server service system: the effect of ξ-choice. Eur. J. Oper. Res. **294**(1), 219–235 (2021)
22. P. Carotenuto, F. Martis, A double dynamic fast algorithm to solve multi-vehicle dial a ride problem. Transp. Res. Procedia **27**, 632–639 (2017)
23. J. Chen, J. Shi, A multi-compartment vehicle routing problem with time windows for urban distribution-a comparison study on particle swarm optimization algorithms. Comput. Ind. Eng. **133**, 95–106 (2019)
24. Z.L. Chen, H. Xu, Dynamic column generation for dynamic vehicle routing with time windows. Transp. Sci. **40**(1), 74–88 (2006)
25. H.K. Chen, C.F. Hsueh, M.S. Chang, The real-time time-dependent vehicle routing problem. Transp. Res. Part E Logist. Transp. Rev. **42**(5), 383–408 (2006)
26. D. Chen, F. Zou, R. Lu, X. Wang, A hybrid fuzzy inference prediction strategy for dynamic multi-objective optimization. Swarm Evolut. Comput. **43**, 147–165 (2018)
27. S. Chen, R. Chen, G.G. Wang, J. Gao, A.K. Sangaiah, An adaptive large neighborhood search heuristic for dynamic vehicle routing problems. Comput. Electr. Eng. **67**, 596–607 (2018)
28. X. Chen, J. Xue, Z. Lei, X. Qian, S.V. Ukkusuri, Online eco-routing for electric vehicles using combinatorial multi-armed bandit with estimated covariance. Transp. Res. Part D Transp. Environ. **111**, 103447 (2022)
29. B.K.S. Cheung, K. Choy, C.L. Li, W. Shi, J. Tang, Dynamic routing model and solution methods for fleet management with mobile technologies. Int. J. Prod. Econ. **113**(2), 694–705 (2008). Special Section on Advanced Modeling and Innovative Design of Supply Chain
30. N. Christofides, The vehicle routing problem, in *Combinatorial Optimization* (Wiley, Hoboken, 1979), pp. 315–338
31. G. Clarke, J.W. Wright, Scheduling of vehicles from a central depot to a number of delivery points. Oper. Res. **12**(4), 568–581 (1964)

32. G.B. Dantzig, J.H. Ramser, The truck dispatching problem. Manag. Sci. **6**(1), 80–91 (1959)
33. J. de Armas, B. Melián-Batista, Variable neighborhood search for a dynamic rich vehicle routing problem with time windows. Comput. Ind. Eng. **85**, 120–131 (2015)
34. Duan, L., Wei, Y., Zhang, J., Xia, Y.: Centralized and decentralized autonomous dispatching strategy for dynamic autonomous taxi operation in hybrid request mode. Transp. Res. Part C Emerg. Technol. **111**, 397–420 (2020)
35. R. Elshaer, H. Awad, A taxonomic review of metaheuristic algorithms for solving the vehicle routing problem and its variants. Comput. Ind. Eng. **140**, 106242 (2020)
36. J. Euchi, A. Yassine, H. Chabchoub, The dynamic vehicle routing problem: solution with hybrid metaheuristic approach. Swarm Evolut. Comput. **21**, 41–53 (2015)
37. A. Fabri, P. Recht, On dynamic pickup and delivery vehicle routing with several time windows and waiting times. Transp. Res. Part B Methodol. **40**(4), 335–350 (2006)
38. H. Fan, Y. Zhang, P. Tian, Y. Lv, H. Fan, Time-dependent multi-depot green vehicle routing problem with time windows considering temporal-spatial distance. Comput. Oper. Res. **129**, 105211 (2021)
39. F. Ferrucci, S. Bock, Real-time control of express pickup and delivery processes in a dynamic environment. Transp. Res. Part B Methodol. **63**, 1–14 (2014)
40. M.L. Fisher, Optimal solution of vehicle routing problems using minimum k-trees. Oper. Res. **42**(4), 626–642 (1994)
41. J.C. Fonseca-Galindo, G. de Castro Surita, J.M. Neto, C.L. de Castro, A.P. Lemos, A multi-agent system for solving the dynamic capacitated vehicle routing problem with stochastic customers using trajectory data mining. Expert Syst. Appl. **195**, 116602 (2022)
42. J.P. Futalef, D. Muñoz-Carpintero, H. Rozas, M.E. Orchard, An online decision-making strategy for routing of electric vehicle fleets. Inf. Sci. **625**, 715–737 (2023)
43. M. Gendreau, F. Guertin, J.Y. Potvin, É. Taillard, Parallel tabu search for real-time vehicle routing and dispatching. Transp. Sci. **33**(4), 381–390 (1999)
44. M. Gendreau, F. Guertin, J.Y. Potvin, R. Séguin, Neighborhood search heuristics for a dynamic vehicle dispatching problem with pick-ups and deliveries. Transp. Res. Part E Logist. Transp. Rev. **14**(3), 157–174 (2006)
45. S.F. Ghannadpour, S. Noori, R. Tavakkoli-Moghaddam, K. Ghoseiri, A multi-objective dynamic vehicle routing problem with fuzzy time windows: model, solution and application. Appl. Soft Comput. **14**, 504–527 (2014)
46. G. Ghiani, F. Guerriero, G. Laporte, R. Musmanno, Real-time vehicle routing: Solution concepts, algorithms and parallel computing strategies. Eur. J. Oper. Res. **151**(1), 1–11 (2003)
47. G. Ghiani, E. Manni, A. Quaranta, C. Triki, Anticipatory algorithms for same-day courier dispatching. Transp. Res. Part E Logist. Transp. Rev. **45**(1), 96–106 (2009)
48. G. Ghiani, A. Manni, E. Manni, A scalable anticipatory policy for the dynamic pickup and delivery problem. Comput. Oper. Res. **147**, 105943 (2022)
49. C.K. Goh, K.C. Tan, A competitive-cooperative coevolutionary paradigm for dynamic multiobjective optimization. Trans. Evol. Comp **13**(1), 103–127 (2009)
50. F. Habibi, F. Barzinpour, S.J. Sadjadi, A multi-objective optimization model for project scheduling with time-varying resource requirements and capacities. J. Ind. Syst. Eng. **10**, 92–118 (2017)
51. A. Haghani, S. Jung, A dynamic vehicle routing problem with time-dependent travel times. Comput. Oper. Res. **32**(11), 2959–2986 (2005)
52. L. Häme, An adaptive insertion algorithm for the single-vehicle dial-a-ride problem with narrow time windows. Eur. J. Oper. Res. **209**(1), 11–22 (2011)
53. S.C. Ho, W.Y. Szeto, Y.H. Kuo, J.M. Leung, M. Petering, T.W. Tou, A survey of dial-a-ride problems: Literature review and recent developments. Transp. Res. Part B Methodol. **111**, 395–421 (2018)
54. L. Hong, An improved LNS algorithm for real-time vehicle routing problem with time windows. Comput. Oper. Res. **39**(2), 151–163 (2012)
55. H. Housroum, T. Hsu, R. Dupas, G. Goncalves, A hybrid GA approach for solving the dynamic vehicle routing problem with time windows, in *2006 2nd International Conference*

on Information & Communication Technologies, vol. 1 (IEEE, Piscataway, 2006), pp. 787–792
56. S. Hua, W. Zeng, X. Liu, M. Qi, Optimality-guaranteed algorithms on the dynamic shared-taxi problem. Transp. Res. Part E Logist. Transp. Rev. **164**, 102809 (2022)
57. M. Hyland, H.S. Mahmassani, Dynamic autonomous vehicle fleet operations: Optimization-based strategies to assign AVs to immediate traveler demand requests. Transp. Res. Part C Emerg. Technol. **92**, 278–297 (2018)
58. S. Ichoua, M. Gendreau, J.Y. Potvin, Vehicle dispatching with time-dependent travel times. Eur. J. Oper. Res. **144**(2), 379–396 (2003)
59. J.J. Jaw, A.R. Odoni, H.N. Psaraftis, N.H. Wilson, A heuristic algorithm for the multi-vehicle advance request dial-a-ride problem with time windows. Transp. Res. Part B Methodol. **20**(3), 243–257 (1986)
60. Y. Jin, X. Ge, L. Zhang, J. Ren, A two-stage algorithm for bi-objective logistics model of cash-in-transit vehicle routing problems with economic and environmental optimization based on real-time traffic data. J. Ind. Inf. Integr. **26**, 100273 (2022)
61. N. Jozefowiez, F. Semet, E.G. Talbi, *From Single-Objective to Multi-Objective Vehicle Routing Problems: Motivations, Case Studies, and Methods* (Springer US, Boston, 2008), pp. 445–471
62. Q. Jun, J. Wang, B.J. Zheng, A hybrid multi-objective algorithm for dynamic vehicle routing problems, in ed. by M. Bubak, G.D. van Albada, J. Dongarra, P.M.A. Sloot, *Computational Science – ICCS 2008* (Springer, Berlin, 2008), pp. 674–681
63. F. Karami, W. Vancroonenburg, G. Vanden Berghe, A periodic optimization approach to dynamic pickup and delivery problems with time windows. J. Sched. **23**, 711–731 (2020)
64. Y. Kergosien, C. Lente, D. Piton, J.C. Billaut, A tabu search heuristic for the dynamic transportation of patients between care units. Eur. J. Oper. Res. **214**(2), 442–452 (2011)
65. M.R. Khouadjia, B. Sarasola, E. Alba, L. Jourdan, E.G. Talbi, A comparative study between dynamic adapted PSO and VNS for the vehicle routing problem with dynamic requests. Appl. Soft Comput. **12**(4), 1426–1439 (2012)
66. P. Kilby, P. Prosser, P. Shaw, Dynamic VRPs: A study of scenarios. University of Strathclyde Technical Report (1998), pp. 1–11
67. K. Knyazkov, I. Derevitsky, L. Mednikov, A. Yakovlev, Evaluation of dynamic ambulance routing for the transportation of patients with acute coronary syndrome in Saint Petersburg. Procedia Comput. Sci. **66**, 419–428 (2015)
68. Ç. Koç, G. Laporte, İ. Tükenmez, A review of vehicle routing with simultaneous pickup and delivery. Comput. Oper. Res. **122**, 104987 (2020)
69. S. Krichen, S.B. Jouida, *Supply Chain Management and Its Applications in Computer Science* (Wiley, Hoboken, 2015)
70. R. Kuo, B. Wibowo, F. Zulvia, Application of a fuzzy ant colony system to solve the dynamic vehicle routing problem with uncertain service time. Appl. Math. Modell. **40**(23–24), 9990–10001 (2016)
71. A. Larsen, The dynamic vehicle routing problem. Ph.D. Thesis, Department of Mathematical Modelling, The Technical University of Denmark, Building 321, DTU, DK-2800 Kgs. Lyngby, 2000
72. A. Larsen, O.B. Madsen, M.M. Solomon, Recent developments in dynamic vehicle routing systems. Vehic. Rout. Problem Latest Adv. New Chall. **43**, 199–218 (2008)
73. E. Lee, X. Cen, H.K. Lo, Scheduling zonal-based flexible bus service under dynamic stochastic demand and time-dependent travel time. Transp. Res. Part E Logist. Transp. Rev. **168**, 102931 (2022)
74. M.W. Levin, K.M. Kockelman, S.D. Boyles, T. Li, A general framework for modeling shared autonomous vehicles with dynamic network-loading and dynamic ride-sharing application. Comput. Environ. Urban Syst. **64**, 373–383 (2017)
75. L. Li, T. Pantelidis, J.Y. Chow, S.E. Jabari, A real-time dispatching strategy for shared automated electric vehicles with performance guarantees. Transp. Res. Part E Logist. Transp. Rev. **152**, 102392 (2021)

76. X. Liang, G.H. de Almeida Correia, K. An, B. van Arem, Automated taxis' dial-a-ride problem with ride-sharing considering congestion-based dynamic travel times. Transp. Res. Part C Emerg. Technol. **112**, 260–281 (2020)
77. C. Lin, K. Choy, G. Ho, H. Lam, G.K. Pang, K. Chin, A decision support system for optimizing dynamic courier routing operations. Expert Syst. Appl. **41**(15), 6917–6933 (2014)
78. X. Liu, J. Luo, A dynamic multi-objective optimization model with interactivity and uncertainty for real-time reservoir flood control operation. Appl. Math. Modell. **74**, 606–620 (2019)
79. Z. Liu, T. Miwa, W. Zeng, M.G. Bell, T. Morikawa, Dynamic shared autonomous taxi system considering on-time arrival reliability. Transp. Res. Part C Emerg. Technol. **103**, 281–297 (2019)
80. A. Lois, A. Ziliaskopoulos, Online algorithm for dynamic dial a ride problem and its metrics. Transp. Res. Procedia **24**, 377–384 (2017)
81. D. Lu, F. Gzara, The robust vehicle routing problem with time windows: Solution by branch and price and cut. Eur. J. Oper. Res. **275**(3), 925–938 (2019)
82. C. Lu, Y. Wu, S. Yu, A sample average approximation approach for the stochastic dial-a-ride problem on a multigraph with user satisfaction. Eur. J. Oper. Res. **302**(3), 1031–1044 (2022)
83. K. Lund, O.B. Madsen, J.M. Rygaard, Vehicle routing problems with varying degrees of dynamism. IMM Institute of Mathematical Modelling (1996)
84. W. Ma, L. Zeng, K. An, Dynamic vehicle routing problem for flexible buses considering stochastic requests. Transp. Res. Part C Emerg. Technol. **148**, 104030 (2023)
85. O.B. Madsen, H.F. Ravn, J.M. Rygaard, A heuristic algorithm for a dial-a-ride problem with time windows, multiple capacities, and multiple objectives. Ann. Oper. Res. **60**(1), 193–208 (1995)
86. J. Mańdziuk, A. Żychowski, A memetic approach to vehicle routing problem with dynamic requests. Appl. Soft Comput. **48**, 522–534 (2016)
87. J.T. Margolis, Y. Song, S.J. Mason, A Markov decision process model on dynamic routing for target surveillance. Comput. Oper. Res. **141**, 105699 (2022)
88. M. Mavrovouniotis, C. Li, S. Yang, A survey of swarm intelligence for dynamic optimization: Algorithms and applications. Swarm Evol. Comput. **33**, 1–17 (2017)
89. K. Meesuptaweekoon, P. Chaovalitwongse, Dynamic vehicle routing problem with multiple depots. Eng. J. **18**(4), 135–149 (2014)
90. S. Mitrović-Minić, R. Krishnamurti, G. Laporte, Double-horizon based heuristics for the dynamic pickup and delivery problem with time windows. Transp. Res. Part B Methodol. **38**(8), 669–685 (2004)
91. J.C. Molina, J.L. Salmeron, I. Eguia, An ACS-based memetic algorithm for the heterogeneous vehicle routing problem with time windows. Expert Syst. Appl. **157**, 113379 (2020)
92. R. Montemanni, L.M. Gambardella, A.E. Rizzoli, A.V. Donati, Ant colony system for a dynamic vehicle routing problem. J. Comb. Optim. **10**(4), 327–343 (2005)
93. T.T. Nguyen, S. Yang, J. Branke, Evolutionary dynamic optimization: a survey of the state of the art. Swarm Evol. Comput. **6**, 1–24 (2012)
94. M. Okulewicz, J. Mańdziuk, The impact of particular components of the PSO-based algorithm solving the dynamic vehicle routing problem. Appl. Soft Comput. **58**, 586–604 (2017)
95. M. Okulewicz, J. Mańdziuk, A metaheuristic approach to solve dynamic vehicle routing problem in continuous search space. Swarm Evol. Comput. **48**, 44–61 (2019)
96. M. Ostermeier, T. Henke, A. Hübner, G. Wäscher, Multi-compartment vehicle routing problems: State-of-the-art, modeling framework and future directions. Eur. J. Oper. Res. **292**(3), 799–817 (2021)
97. N. Ouertani, H.B. Ramdhan, S. Krichen, I. Nouaouri, H. Allaoui, A new evolutionary method to deal with the dynamic vehicle routing problem, in *2018 IEEE International Conference on Technology Management, Operations and Decisions (ICTMOD)* (IEEE, Piscataway, 2018), pp. 1–5
98. N. Ouertani, H. Ben-Romdhane, S. Krichen, A decision support system for the dynamic hazardous materials vehicle routing problem. Oper. Res. **22**, 1–26 (2020)

99. N. Ouertani, H. Ben-Romdhane, S. Krichen, I. Nouaouri, A vector evaluated evolutionary algorithm with exploitation reinforcement for the dynamic pollution routing problem. J. Comb. Optim. **44**(2), 1011–1038 (2022)
100. A. Pessoa, R. Sadykov, E. Uchoa, Enhanced branch-cut-and-price algorithm for heterogeneous fleet vehicle routing problems. Eur. J. Oper. Res. **270**(2), 530–543 (2018)
101. J. Pierotti, J.T. van Essen, MILP models for the dial-a-ride problem with transfers. EURO J. Transp. Logist. **10**, 100037 (2021)
102. V. Pillac, M. Gendreau, C. Guéret, A.L. Medaglia, A review of dynamic vehicle routing problems. Eur. J. Oper. Res. **225**(1), 1–11 (2013)
103. H.N. Psaraftis, Dynamic vehicle routing problems. Vehicl. Rout. Methods Stud. **16**, 223–248 (1988)
104. H.N. Psaraftis, Dynamic vehicle routing: status and prospects. Ann. Oper. Res. **61**(1), 143–164 (1995)
105. H.N. Psaraftis, M. Wen, C.A. Kontovas, Dynamic vehicle routing problems: three decades and counting. Networks **67**(1), 3–31 (2016)
106. M. Rabbani, R. Heidari, R. Yazdanparast, A stochastic multi-period industrial hazardous waste location-routing problem: integrating NSGA-II and Monte Carlo simulation. Eur. J. Oper. Res. **272**(3), 945–961 (2019)
107. M. Ramezani, M. Nourinejad, Dynamic modeling and control of taxi services in large-scale urban networks: a macroscopic approach. Transp. Res. Part C Emerg. Technol. **94**, 203–219 (2018)
108. M. Riedler, G. Raidl, Solving a selective dial-a-ride problem with logic-based benders decomposition. Comput. Oper. Res. **96**, 30–54 (2018)
109. B.H.O. Rios, E.C. Xavier, F.K. Miyazawa, P. Amorim, E. Curcio, M.J. Santos, Recent dynamic vehicle routing problems: a survey. Comput. Ind. Eng. **160**, 107604 (2021)
110. U. Ritzinger, J. Puchinger, R.F. Hartl, A survey on dynamic and stochastic vehicle routing problems. Int. J. Prod. Res. **54**(1), 215–231 (2016)
111. U. Ritzinger, J. Puchinger, C. Rudloff, R.F. Hartl, Comparison of anticipatory algorithms for a dial-a-ride problem. Eur. J. Oper. Res. **301**(2), 591–608 (2022)
112. N.R. Sabar, A. Bhaskar, E. Chung, A. Turky, A. Song, A self-adaptive evolutionary algorithm for dynamic vehicle routing problems with traffic congestion. Swarm Evol. Comput. **44**, 1018–1027 (2019)
113. D.O. Santos, E.C. Xavier, Taxi and ride sharing: A dynamic dial-a-ride problem with money as an incentive. Expert Syst. Appl. **42**(19), 6728–6737 (2015)
114. B. Sarasola, K.F. Doerner, V. Schmid, E. Alba, Variable neighborhood search for the stochastic and dynamic vehicle routing problem. Ann. Oper. Res. **236**(2), 425–461 (2016)
115. M. Schyns, An ant colony system for responsive dynamic vehicle routing. Eur. J. Oper. Res. **245**(3), 704–718 (2015)
116. H. Seyyedhasani, J.S. Dvorak, Dynamic rerouting of a fleet of vehicles in agricultural operations through a dynamic multiple depot vehicle routing problem representation. Biosyst. Eng. **171**, 63–77 (2018)
117. S. Shahparvari, B. Abbasi, Robust stochastic vehicle routing and scheduling for bushfire emergency evacuation: an Australian case study. Transp. Res. Part A Policy Pract. **104**, 32–49 (2017)
118. P.K. Sheridan, E. Gluck, Q. Guan, T. Pickles, B. Balcıog, B. Benhabib et al., The dynamic nearest neighbor policy for the multi-vehicle pick-up and delivery problem. Transp. Res. Part A Policy Pract. **49**, 178–194 (2013)
119. K. Shin, S. Han, A centroid-based heuristic algorithm for the capacitated vehicle routing problem. Comput. Inf. **30**(4), 721–732 (2011)
120. N. Soeffker, M.W. Ulmer, D.C. Mattfeld, Stochastic dynamic vehicle routing in the light of prescriptive analytics: a review. Eur. J. Oper. Res. **298**(3), 801–820 (2022)
121. M.M. Solomon, Algorithms for the vehicle routing and scheduling problems with time window constraints. Oper. Res. **35**(2), 254–265 (1987)

122. Z. Steever, M. Karwan, C. Murray, Dynamic courier routing for a food delivery service. Comput. Oper. Res. **107**, 173–188 (2019)
123. A. Strauss, N. Gülpınar, Y. Zheng, Dynamic pricing of flexible time slots for attended home delivery. Eur. J. Oper. Res. **294**(3), 1022–1041 (2021)
124. A. Tafreshian, M. Abdolmaleki, N. Masoud, H. Wang, Proactive shuttle dispatching in large-scale dynamic dial-a-ride systems. Transp. Res. Part B Methodol. **150**, 227–259 (2021)
125. É.D. Taillard, A heuristic column generation method for the heterogeneous fleet VRP. RAIRO-Oper. Res. **33**(1), 1–14 (1999)
126. K. Tan, T. Lee, E. Khor, Evolutionary algorithms for multi-objective optimization: Performance assessments and comparisons. Artif. Intell. Rev. **17**(4), 251–290 (2002)
127. H. Tang, M. Hu, Dynamic vehicle routing problem with multiple objectives: Solution framework and computational experiments. Transp. Res. Record **1923**(1), 199–207 (2005)
128. X. Tang, J. Yang, X. Lin, F. He, J. Si, Dynamic operations of an integrated mobility service system of fixed-route transits and flexible electric buses. Transp. Res. Part E Logist. Transp. Rev. **173**, 103081 (2023)
129. V. Torretta, E.C. Rada, M. Schiavon, P. Viotti, Decision support systems for assessing risks involved in transporting hazardous materials: a review. Safety Sci. **92**, 1–9 (2017)
130. P. Toth, D. Vigo, *Vehicle Routing: Problems, Methods, and Applications*. MOS-SIAM Series on Optimization, 2nd edn. (Society for Industrial and Applied Mathematics (SIAM), Philadelphia, 2014)
131. M.W. Ulmer, N. Soeffker, D.C. Mattfeld, Value function approximation for dynamic multi-period vehicle routing. Eur. J. Oper. Res. **269**(3), 883–899 (2018)
132. M.W. Ulmer, J.C. Goodson, D.C. Mattfeld, B.W. Thomas, On modeling stochastic dynamic vehicle routing problems. Eur. J. Transp. Logist. **9**(2), 100008 (2020)
133. S. Vonolfen, M. Affenzeller, Distribution of waiting time for dynamic pickup and delivery problems. Ann. Oper. Res. **236**, 359–382 (2016)
134. VOSviewer: Welcome to vosviewer. VOSviewer visualizing scientific landscapes (2021)
135. Y. Wang, W. Szeto, The dynamic bike repositioning problem with battery electric vehicles and multiple charging technologies. Transp. Res. Part C Emerg. Technol. **131**, 103327 (2021)
136. F. Wang, F. Liao, Y. Li, X. Yan, X. Chen, An ensemble learning based multi-objective evolutionary algorithm for the dynamic vehicle routing problem with time windows. Comput. Ind. Eng. **154**, 107131 (2021)
137. Z. Wang, K. Ye, M. Jiang, J. Yao, N.N. Xiong, G.G. Yen, Solving hybrid charging strategy electric vehicle based dynamic routing problem via evolutionary multi-objective optimization. Swarm Evol. Comput. **68**, 100975 (2022)
138. S. Wohlgemuth, R. Oloruntoba, U. Clausen, Dynamic vehicle routing with anticipation in disaster relief. Socio-Econ. Plann. Sci. **46**(4), 261–271 (2012)
139. Y. Wu, L. Shi, X. Liu, A new dynamic strategy for dynamic multi-objective optimization. Inf. Sci. **529**, 116–131 (2020)
140. X. Xiang, J. Qiu, J. Xiao, X. Zhang, Demand coverage diversity based ant colony optimization for dynamic vehicle routing problems. Eng. Appl. Artif. Intell. **91**, 103582 (2020)
141. J. Yang, P. Jaillet, H. Mahmassani, Real-time multivehicle truckload pickup and delivery problems. Transp. Sci. **38**(2), 135–148 (2004)
142. Z. Yang, J.P. van Osta, B. van Veen, R. van Krevelen, R. van Klaveren, A. Stam, J. Kok, T. Bäck, M. Emmerich, Dynamic vehicle routing with time windows in theory and practice. Nat. Comput. **16**(1), 119–134 (2017)
143. S. Yoon, L.A. Albert, A dynamic ambulance routing model with multiple response. Transp. Res. Part E Logist. Transp. Rev. **133**, 101807 (2020)
144. X. Zhan, W. Szeto, X.M. Chen, The dynamic ride-hailing sharing problem with multiple vehicle types and user classes. Transp. Res. Part E Logist. Transp. Rev. **168**, 102891 (2022)
145. H. Zhang, Q. Zhang, L. Ma, Z. Zhang, Y. Liu, A hybrid ant colony optimization algorithm for a multi-objective vehicle routing problem with flexible time windows. Inf. Sci. **490**, 166–190 (2019)

146. S. Zhang, W. Zhang, Y. Gajpal, S.S. Appadoo, *Ant Colony Algorithm for Routing Alternate Fuel Vehicles in Multi-depot Vehicle Routing Problem* (Springer, Singapore, 2019), pp. 251–260
147. X. Zhang, J. Zhang, X. Fan, Offline approximate value iteration for dynamic solutions to the multivehicle routing problem with stochastic demand. Comput. Oper. Res. **146**, 105884 (2022)
148. J. Zhang, K. Luo, A.M. Florio, T. Van Woensel, Solving large-scale dynamic vehicle routing problems with stochastic requests. Eur. J. Oper. Res. **306**(2), 596–614 (2023)
149. T. Zhou, M.L. Kris, D. Creighton, C. Wu, GMIX: graph-based spatial-temporal multi-agent reinforcement learning for dynamic electric vehicle dispatching system. Transp. Res. Part C Emerg. Technol. **144**, 103886 (2022)

Multi-Objective Optimization for Electric Vehicle Routing Problem: Literature Review

Anouar Haddad, Takwa Tlili, and Saoussen Krichen

1 Introduction

Optimization refers to problems that are in need of hunting for the best configuration of a set of variables in order to accomplish a specific intention. Basically, in recent years, a variety of problems labeled combinatorial optimization problems have appeared and become more and more challenging due to their high complexity and their requirements for highly efficient methods to resolve them. Its multitudinous applications in both theoretical and experimental domains have persuaded combinatorial optimization. Among these areas, notable mentions include logistics, transportation, distribution, and delivery, which take advantage of the impressive advancement in scientific research in combinatorial optimization to deal especially with vehicle routing problems and to improve the formulation of these problems as well as the analysis and the application of their solution processes. From a practical standpoint, efficient scheduling of vehicles can help businesses boost productivity by facilitating long-term planning and managing resource limitations effectively. Over time, and due to its valuable impact, the vehicle routing problem has been swept away and many variants have appeared. Among its variants is the Electric Vehicle Routing Problem (EVRP), which aims to determine the optimal routes and schedules for a fleet of electric vehicles to serve a set of customers, considering specific constraints and characteristics of electric vehicles (EVs) such as battery range, charging infrastructure, and energy consumption. The EVRP caught our attention because it is not only looking for economic growth but also for environmental sustainability, especially when it aims to improve more than

A. Haddad (✉) · T. Tlili · S. Krichen
LARODEC, Institut Superieur de Gestion de Tunis, Tunis, Tunisia
e-mail: saoussen.krichen@isg.rnu.tn

I. Alharbi et al. (eds.), *Advances in Computational Logistics and Supply Chain Analytics*, Unsupervised and Semi-Supervised Learning,
https://doi.org/10.1007/978-3-031-50036-7_2

one objective function. Hither, multi-objective optimization takes place to deal specifically with multiple conflicting objective functions.

In this paper, the fundamental approaches of multi-objective optimization are explored, placing particular emphasis on their application in the context of the Electric Vehicle Routing Problem (EVRP). The chapter unfolds as follows:

1. Introduction: Sect. 1 provides an overview of the general context, setting the stage for the subsequent discussions. It introduces the motivation behind multi-objective optimization and its relevance in solving complex optimization problems.
2. Approaches to Multi-Objective Optimization: Sect. 2 delves into various approaches employed in multi-objective optimization. It highlights key methodologies, such as Pareto-based algorithms, evolutionary algorithms, and meta-heuristic techniques. These approaches play a crucial role in tackling the inherent trade-offs and complexities associated with multi-objective problems.
3. EVRP Overview: Sect. 3 offers a comprehensive understanding of the Electric Vehicle Routing Problem (EVRP), its various variants, and policies. It outlines the specific challenges and considerations that arise in the domain of EVRP.
4. Specifying Multi-Objective Problems with EVs: Sect. 4 focuses on the specification and formulation of multi-objective optimization problems in the context of Electric Vehicles (EVs). It highlights the unique characteristics and constraints associated with EVs, such as range limitations, charging infrastructure, and energy consumption.
5. State-of-the-Art Methods for EVRP: Sect. 5 delves into a comprehensive discussion of different state-of-the-art methods proposed in the literature to address the EVRP. It examines and analyzes various optimization algorithms, heuristics, and solution techniques specifically tailored for EVRP, considering multiple objectives.
6. Future Research Directions: Sect. 6 sheds light on future research directions within the domain of multi-objective optimization for EVRP. It discusses potential areas of exploration, such as incorporating real-time data, integrating renewable energy sources, or incorporating uncertainty and dynamic aspects into the optimization models.
7. Conclusion: Sect. 7 presents the conclusion of the chapter, summarizing the key findings and insights. It underscores the significance of multi-objective optimization in addressing EVRP challenges and emphasizes the need for further advancements in methodologies and techniques.

2 Multi-Objective Optimization

Since many dangers threaten our surroundings, people are giving more frequent heed to energy conservation, ecology, green development, gain maximization, cost minimization, etc. in distinct fields. For this, optimization methods are competent and apt for dealing with real-world issues. Optimization problems are one of great interest in scientific research. These problems may be split into two categories based on the number of objective functions that each problem has: single-objective optimization problems and multi-objective optimization problems. Multi-objective optimization, also called multi-criteria optimization or vector optimization, is a branch of mathematical optimization where certain conflicting objectives have to be optimized simultaneously over an attainable set determined by constraint functions [1]. It deals specifically with optimization problems possessing multiple objective functions and it belongs to finding the best solution values corresponding to more than one desired intention. The purpose of using multi-objective optimization is because it does not require a sophisticated formulation of the problem rather than simple equations, which consequently cuts down the problem [2]. The mathematical equation of a multi-objective optimization problem can be formulated as follows:

$$\min / \max \qquad f_1(x), f_2(x), \ldots, f_n(x) \qquad \text{with } x \in S \tag{1}$$

where $n \geq 2$ is the number of objective functions, x is a solution, S is the feasible solution set, $f_n(x)$ is the nth objective function, and min/max is the decision rule operator.

To solve the multi-objective optimization problem, two distinct approaches can be exploited, namely the method using multi-objective optimization algorithms based on Pareto optimality and the classic scalar method using single-objective optimization algorithms.

The various objective functions are about to be flattened into a single one via the scalar approach. These functions consist generally of three types: scalarization method, goal programming method, and ε-constraint method [3].

- **The Scalarization Method:** It is also known as the weighted sum. It combines all objectives into one single-objective function by multiplying the weight and merging the scalar function in a fitness function [2]. The following equation illustrates how the scalarization method merges multi-objective functions into the scalar fitness function.

$$\min / \max \qquad \sum_{i=1}^{n} w_i f_i(x) \qquad \text{with } x \in S \tag{2}$$

  where n is the number of objective functions, x is a solution, S is the feasible solution set, and w is the weight. Weights can be allocated in various ways, including equal, subjective, objective, and a mix of objective and subjective weights. The allocation of weights has to be completed before the optimization process for its critical determination because the outcomes are highly dependent on the weights assigned and principally determined by them [4].
- **Goal Programming Method:** It is a distinct scalar approach that does not refer directly to the decision rule (minimization or maximization) of all the objectives. So, in this method, a goal is carefully specified, and for each aim, a unique target value–typically numeric–is formed. The search has led to reducing the gap

between the existing solution and the goal such that these two entities are close enough to one another [5].

- **ε-Constraint Method:** The last approach found in this survey of the literature is the ε-constraint method where only one objective is kept to be optimized and the others are restricted to be considered as constraints.
- **Pareto method:** This method uses the concept of domination, where it discriminates between the dominated solutions and the non-dominated ones using performance indicators. So, it generates a set of all-efficient solutions called the Pareto front for the purpose of delimiting the spotlight to this set of preferences and making a compromise solution called a trade-off within this set, instead of considering the full range. Two distinct terms in the Pareto optimum solution need to be recorded when using the Pareto approach [2]. These are the terms:
  - *Anchor Point*: It is captured when the objective function reaches the best.
  - *Utopia Point*: It is the intersection of an objective function's maximum and minimum values with those of another objective function.

The different approaches cited previously can be solved either under determinism considering antagonistic objectives to find efficient solutions or else under uncertainty regarding whether some facts are ambiguous and whose effects are unknown [6].

3 Electric Vehicle Routing Problem

During the last decade, electric vehicles have become a new concept in the world of the automotive industry, which is growing in popularity day after day and receiving great attention not only from human society but also from governments, in reason of their superb contribution toward a vigorous and durable environment, such as in the reduction of toxic gas emissions, a decrease of the dependence on fossil fuel, high energy efficiency, low noise pollution, etc.

The EVRP introduces specific challenges and considerations that differentiate it from conventional VRPs. These differences arise due to the unique characteristics and constraints associated with EVs. One of the primary challenges in the EVRP is the limited range of EVs compared to conventional vehicles. EVs typically have a finite battery capacity, and their range is influenced by factors such as battery technology, weather condition, and driving behavior [7]. In conventional VRP, the focus is primarily on minimizing distances traveled or optimizing time-based metrics. However, in the EVRP, the objective expands to consider minimizing energy consumption or maximizing the number of customers served while ensuring sufficient battery capacity for the entire route. Another key difference in the EVRP is the inclusion of charging infrastructure as a critical factor. Unlike conventional VRPs that focus on capacity constraints, the EVRP introduces an additional constraint related to the charging policy or the time required at each station. The existence of these new optimized variables and objective functions that influence

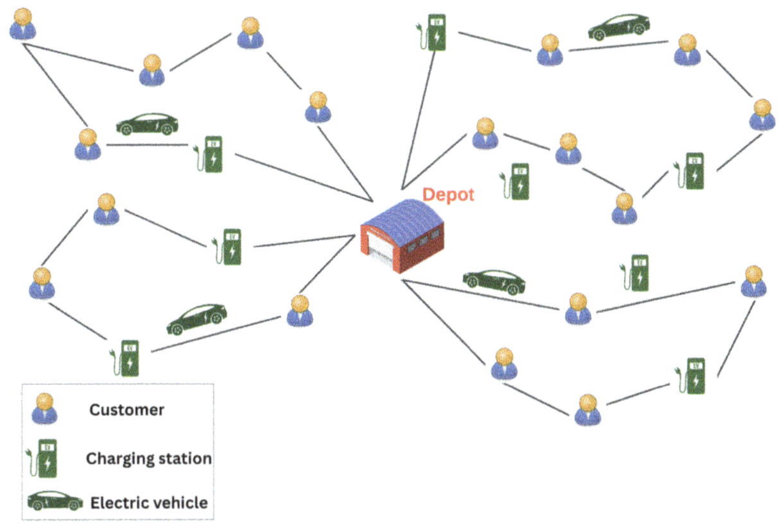

Fig. 1 An illustration of the EVRP solution with optimized routes and charging stations

the decision-making process in route planning, vehicle scheduling, and charging station selection does contribute to the mathematical differentiation of EVRP from VRP. However, the EVRP can still be represented as a vehicle routing problem with additional constraints and objectives related to electric vehicle charging.

The EVRP like any VRP has been extensively studied, resulting in a wide range of algorithms and optimization techniques such as exact methods, heuristics, and meta-heuristics which offer robust solutions for various problem instances. However, it is important to note that the efficiency of existing techniques can vary depending on the specific problem instance, its size, and the constraints involved.

3.1 EVRP Variants

Afterward, its introduction by Dantzig and Ramser [8], the Vehicle Routing Problem (VRP) and its variants have been extensively reviewed and many researchers have worked on this subject, thereby plentiful scientific papers can be found in the literature. The basic VRP strives to determine the optimal routes starting and ending at the depot by visiting a set of customers through multiple vehicles while respecting some constraints. While the electric vehicle routing problem is an extension of the traditional VRP, which differs by finding a set of vehicle route plans for EVs that satisfies a given objective function while complying with several limitations and operational strategies for it [7]. Figure 1 illustrates the EVRP:

A wide range of EVRP variants have been examined and researchers began adapting them for the EVRP context as soon as EVs became commonplace. Some new problem-specific versions evolved due to EV routing's specific features or constraints such as the number of depots, time of service, nature of demands, vehicle's characteristics, type and size of the fleet, energy and charging policies.

3.1.1 Electric Traveling Salesman Problem

The ETSP is one of the simplest forms of EVRP. It is an exception that extends TSP by simply taking into account only one EV. The latter may be recharged at the station while the tour is in progress. The main purpose is to identify the shortest Hamiltonian circuit of a group of customers while guaranteeing that the battery level is always positive. Another variation known as ETSP with time windows is explored when each consumer is subjected to a time window restriction.

3.1.2 EVRP with Time Windows

In this variant, the fleet is solely made up of EVs and customers have time windows within which deliveries should be done. EVRPTW is widely studied in the literature, so Keskin et al. [9] has solved this problem with an adaptive large neighborhood search. Also, Cortés-Murcia et al. [10] proposed a mathematical model as well as an iterated local search meta-heuristic to solve the EVRPTW with partial recharges and satellite customers.

3.1.3 Two-Echelon EVRP

The Two-Echelon EVRP (2E-EVRP) seeks the least-cost delivery routes to transport goods from the central depot to the satellite facilities with conventional vehicles in the first echelon, and from the satellites to the customers using an electric fleet of vehicles in the second echelon. Researchers such as Breunig et al. [11] have introduced the E2-EVRP as a prototypical problem and discussed its impact on the optimized battery-powered distribution systems.

3.1.4 EVRP with Battery Swapping Stations

The EVRP with Battery Swapping (EVRP-BS) stations considers a modern charging policy which is the battery swap instead of the former one and also takes into account a specific kind of EVS which is battery electric vehicles. So, in this variant, the BEV visits the station to exchange the exhausted battery with a fully charged one. It should be noted that numerous variants are not cited in the aforementioned context. For example, Kyriakakis et al. [12] represented the EVRP with drones to minimize the energy consumption for aerial deliveries. While Bahrami et al. [13] suggested a plugin hybrid EVRP for energy minimization. Without neglecting the multi-depot EVRP, EVRP with a mixed fleet, EVRP with linear/nonlinear charging function, electric pickup and delivery problems and many other variants are obviously included in this challenge.

3.2 Energy Sources

Batteries, fuel cells, capacitors, and hybrid storage systems are among the practical energy sources being suggested for EVs, where each energy storage system has its own features [14]. All of them, the batteries and capacitors are energy storage systems in which electrical energy is reserved during charging, whereas the fuel cells are single-handed and can generate energy by themselves.

- Battery: Currently, batteries have been considered the most dominant energy source for EVs [15], because of their technological advancement and affordable cost. An EV battery is a rechargeable battery, considered an electrochemical storage system, which serves to power the electric motors over sustained periods of time. The repetition of this process has a direct effect on the amount of charge the battery can hold. The EV market has been significantly impacted by advancements in battery technology because batteries are the only source of driving power for EVs, whereas the most favored ones are lead-acid batteries and Li-ion-based batteries [14].
- Fuel cell: Over the past several eras, the need for fossil fuels has gradually expanded around the world due to international troubles problems, the reliance on fossil fuels for future energy needs is unsustainable. Contrary to batteries, fuel cells may maintain electrical energy as long as there is a fuel source, therefore virtually having no cycle-life restriction. Also, it should be emphasized that the fuel cell itself is not capable of receiving electricity produced by EVs while braking or going downhill [15]. So driving electric power comes mainly from the fuel cell.
- Capacitor: As the capacitor's capacity has increased, the progress in ultra-capacitors (super-capacitors) has shown promise for use in EVs. An ultra-capacitor is an electromagnetic storage system where electrodes and electrolytes store and release static energy and it can be considered as a battery replacement [14].

  The usage of ultra-capacitors is extremely appealing to supply quick energy for acceleration as well as to accept immediate regenerative energy during braking and downhill, provided that the average energy required for EV application can be given by other energy sources. Also, its specific energy is too low for such an application to be employed as the only energy source for EVs [15].
- Hybrid storage system: A battery and super-capacitor, battery and fuel cell, super-capacitor and fuel cell, or battery super-capacitor and fuel cell make up a hybrid storage system, which is composed of two or more energy storage systems. Every combination has its own benefits, such as high capacity, high specific power, and

high power density. Another combination has a long lifespan, high-temperature tolerance, and low discharge rate [14].

3.3 Charging Strategies

Regardless of locations and types of the recharging stations, the battery capacity of the EVs and also recharging rules are the main two concerns taken into account under various suppositions and expressed as such in the mathematical models in the majority of research connected to the EVRP. The most typical technique for refueling EVs is conductive charging which is used to determine how much of the battery's power can or must be recovered once the car is charged. For these, recharging strategies go under one of two categories: full charging and partial charging [16].

- **Full Charging Strategy:** It implies that the vehicle's battery must be powered up each time it visits a charging station or a specific rechargeable point before it may stop charging.
- **Partial Charging Strategy:** It means that the duration and amount of charge for each recharge operation are related to the next service task of the vehicle so that the EV can leave the charging station at any charge level depending on the time spent charging. Generally, the most frequent and explored charging policies are full and partial charging strategies. However, due to the existence of a separate class of stations from conductive charging stations called battery swapping stations, another charging strategy has appeared: the battery swapping method.
- **Battery Swapping Strategy:** It is an alternate method of recharging EVs. In most cases, it takes only a small amount of time, about five minutes, to replace the vehicle's battery with a brand-new, fully charged one, which is comparable to stopping at a fueling station for conventional vehicles [17]. Hence, at battery swapping stations, the current battery in the EV is replaced on its way to providing service with a fully charged one.
  Logistics and transportation enterprises have set up battery swapping stations and charging stations based on the cutting-edge technology and usage characteristics of EV batteries for the purpose of overcoming certain challenges.

3.4 Vehicles Characteristics

To characterize the vehicles, two types of classification can be considered: the first according to the degree of homogeneity and the second according to the type of the EV system itself. For the first classification, there are two main classes:

- Homogeneous fleet of vehicles: The homogeneous fleet of EVs considers a set of available vehicles with the same characteristics from this the notion of homogeneity comes. This set is defined as a specific capacity Q per vehicle. The homogeneous fleet of EVs is dispatched from a single depot, it starts the routing from the central depot, visits each customer exactly once and at the end it returns to this depot.
- Heterogeneous fleet of vehicles: While most methods have acquired homogeneous fleets, real-world problems in logistics and transportation usually consider multiple terminals and series of vehicles with distinct capacities. A heterogeneous fleet of EVs is the opposite of the homogeneous one which considers vehicles with different capacities and costs, charging technology, etc. The heterogeneous EVRP is used to design the routing of a heterogeneous fleet of EVs in order to service a set of customers each starting and ending at the central depot. It takes into account that the number of vehicles of each type is considered limitless.

While according to the nature of the EV system, three major kinds of EVs are distinguished, namely fully battery EVs, hybrid EVs, and plug-in EVs.

- Fully battery EV (BEV): BEV is a kind of EV that is powered exclusively through electric motors and uses energy stored in rechargeable battery packs, with no alternative source of power.
- Hybrid EV (HEV): HEV is the combination of engine power along with electric power. It is easy to conduct, has an extended driving range, and is generally driven in both rural and urban regions. When it rolls downtown, it boosts completely the battery and switches on the motor so that the battery pack gives the reserve power for driving [14].
- Plug-in electric vehicle (PEV): PEV is any motor vehicle that can be recharged from an external source of electricity to store electrical energy within its onboard rechargeable battery packs which contribute to powering the electric motor and driving the wheels for propulsion. PEV is a subclass of EVs that consists of Battery Electric Vehicles (BEVs) and PEVs.

The previous section was summarized in the Fig. 2 to produce the general framework of the EVRP review. In the scheme, four concerns caught our attention. First of all, the EVRP variants are uncountable and differ either by the charging method or by some practical factors like the time window. The second concern is the different energy sources from which the EV draws its power. Then the charging strategy was mentioned to distinguish between the traditional recharging policy and the in-vogue one. The last heed has been assigned to the vehicles' characteristics which are related either to the nature of the fleet or the vehicle type.

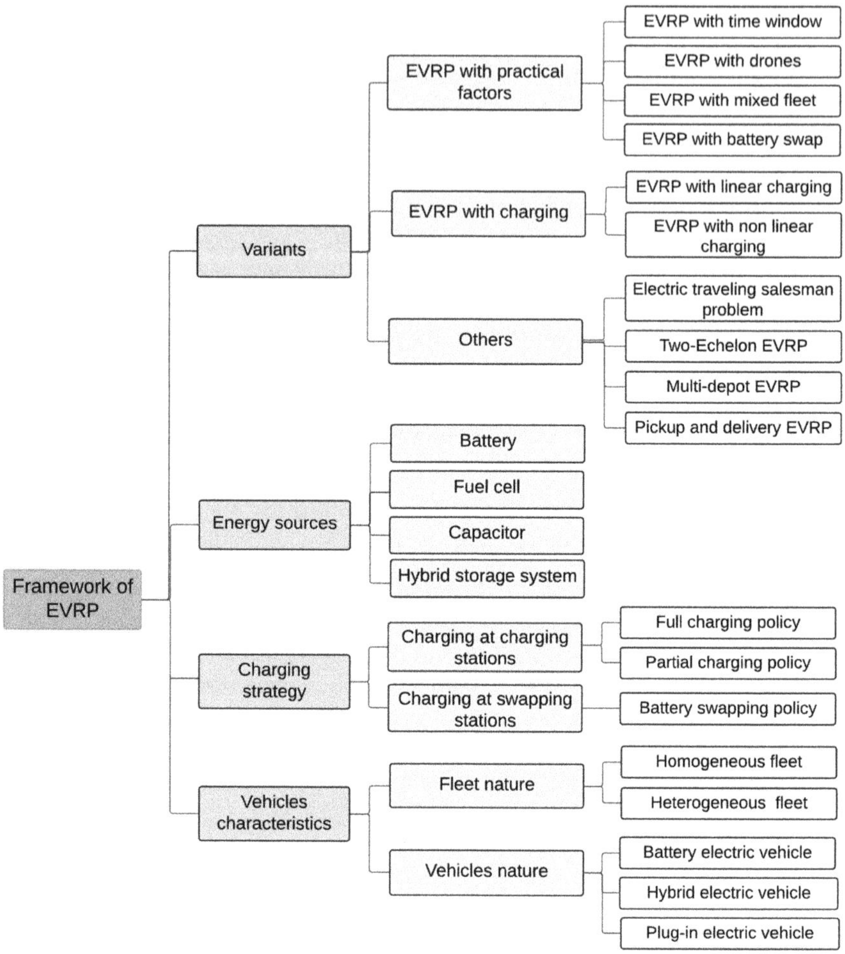

Fig. 2 EVRP framework: variants, energy sources, charging strategies, and vehicles characteristics

4 Multi-Objective Electric Vehicle Routing Problem

The majority of the problems encountered in industry, particularly in logistics and transportation, are multi-objective in nature, but even though these problems are used to model real cases, a common approach is to set them up with the single objective of minimizing the cost or maximizing the total reward. However, the goal might not necessarily be restricted to the cost. In reality, by simply adding new objectives, many more factors such as time, distance, etc. may be considered.

Multi-objective EVRPs (MO-EVRP) are generally used in three ways [18] either to extend classic problems in order to improve their practical application or to generalize classic problems by adding objectives instead of constraints or to study real-life cases like the well-known school-bus route planning [19] which aims to minimize the vehicle purchasing cost, recharging station building cost, environmental cost, and distance cost. Also, without forgetting their application in the field of home health care encountered especially during the spread of the COVID-19 epidemic, here Bahri et al. [6] depict a novel approach to routing problem to minimize the total travel cost for the staff and the waiting time for patients, where the routes are served by EVs.

The tour, node activity, or resources are the three most common objectives that can be classified based on the aspect of the problem to which each objective is linked [18]. Newline Objectives related to the tour consist generally either in minimizing the cost, which is related to an economic criterion like the traveled distance or the required time, or else minimizing balancing objectives like the number of visited customers or the number of delivered goods, or else minimizing the makespan, which is the total time to complete all the tours. While objectives related to node activity involve time windows or the sum of distances or the sum of vehicle stops. And finally, objectives related to resources implicate either goods or vehicles. The EVRP literature reveals the following fundamental objective functions:

- The total distance traveled
- The total time required
- The number of EVs
- The total charging cost
- The energy consumption
- Another operational cost like the waiting cost

Generally, VRP researchers employ the first three objective function aspects. It follows that the majority of EVRP researchers likewise take these objectives into account. The battery and energy consumption intervene as restrictions for the route scheduling in studies that merely take these objectives into account as these two factors have an immediate impact on the routing plan and the total required time. Many research papers incorporate a bunch of basic objective function elements, both from the original VRP and additional energy-related components. The majority of the publications emphasize minimization of the overall traveled distance and it should be noted that the best part of research considers EV energy usage and their batteries' capacity.

5 State-of-the-Art Methods for the MO-EVRP

This section outlines a taxonomic literature review of various existing solution approaches for solving the EVRP. Exact methods, problem-specified heuristics, meta-heuristics, and hybrid approaches are the four categories into which methods for solving optimization problems are classified as shown in Fig. 3. Each has its

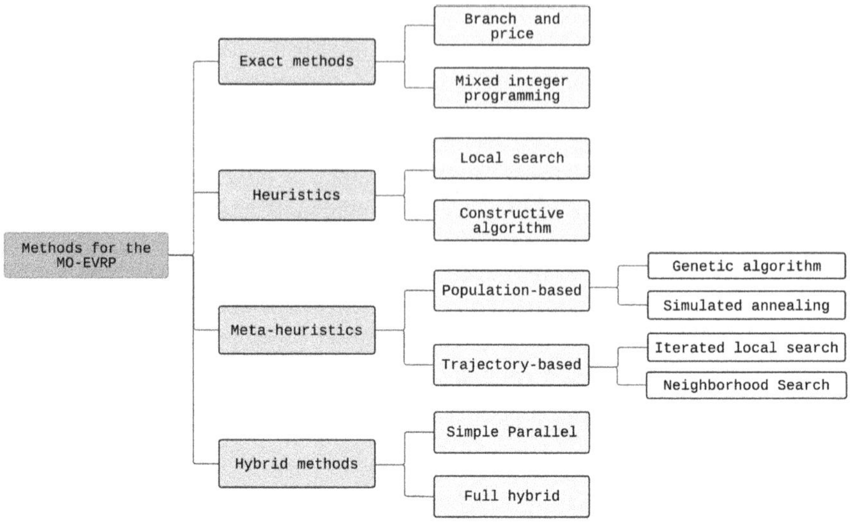

Fig. 3 EVRP solution methods: classification and approaches

own benefits and drawbacks, which must be taken into account when selecting the appropriate approach for a given problem.

5.1 *Exact Methods*

An exact method, also known as an exact algorithm, offers the advantage of providing rigorous guarantees on optimality. By leveraging mathematical principles or sophisticated data structure search techniques, an exact method exhaustively explores the entire solution space to guarantee the discovery of all optimal solutions. In theory, the optimality of the resulting solution can be mathematically proven, instilling confidence in the solution's quality. However, as the number of instances in a problem grows, the computational complexity of exact methods increases exponentially. This exponential growth renders exact methods impractical for large-scale problems, where the solution space becomes overwhelmingly vast. The sheer number of potential solutions and the intricate relationships between variables and constraints make it computationally infeasible to exhaustively search for the optimal solution within a reasonable time frame. The exact methods found in the literature to solve the MO-EVRP are described below:

5.1.1 Branch and Price Method

Branch and price entail a combination of column generation and branch and bound methods to solve integer linear optimization problems. By resolving pricing issues, columns may be added to the linear programming relaxation at each node of the search tree. Branch-and-price-and-cut is another comparable method that incorporates the creation of columns and rows.

Lee [20] has suggested a branch-and-price method for charging networks to resolve the EVRP with nonlinear charging time to optimality. Wu and Zhang [21] have also proposed this method to fix the two-echelon EVRP by using labeling and column generation algorithms.

5.1.2 Mixed Integer Programming

When some variables, but not all, are restricted to having integer values at the optimal solution (i.e. when variables are not all discrete), then mixed integer programming (MIP) is about to take place. However, when all of the decision variables are required to be integers, the contrary is known as a pure integer program. A recent mixed integer programming model for the EVRP was introduced by Ferro et al. [22] to figure out the issue of planning an electrically powered freight transportation service and has demonstrated high-quality solutions for a limited number of customers.

5.2 *Meta-Heuristics*

A meta-heuristic is a powerful and versatile approach that guides heuristics in solving complex optimization problems across various domains. Unlike heuristics, which are tailored to specific problems, meta-heuristics provide a general framework that can be applied to a wide range of optimization problems. More precisely, a meta-heuristic is an iterative masterpiece that generates a single solution or a set of solutions in each loop to optimize a candidate solution with regard to a specific quality metric, hence producing high-quality solutions efficiently. In the following, an overview of some meta-heuristics used for the multi-objective EVRP is given.

5.2.1 Adaptive Large Neighborhood Search

Adaptive Large Neighborhood Search (ALNS) is an extension of Large Neighborhood Search (LNS) where the user may select as many heuristics as he wants. The algorithm will then allocate each heuristic a weight that measures its effectiveness. When alternative options are not available, it is presumed that previous success is the best predictor of future success. ALNS depends on a variety of destruction and repair processes that are chosen based on their previous performance in optimizing the existing solution. The multi-objective EVRP with time windows was solved by dint of the ALNS meta-heuristic by Keskin et al. and has shown good results both in terms of quality and of computational time [9]. Erdelic et al. [23] also tested the ALNS performance using two recharging policies and produced solutions in a reasonable time.

5.2.2 Local Search

Local Search (LS) is a meta-heuristic for tackling computationally hard optimization problems. The fundamental tenet of local search is that it tends to focus on a small segment of the search space. In such a manner, solutions are successfully modified by moves which alter them locally, and neighborhoods that list all surrounding solutions that may be reached with a single move, define valid transformations. This class includes the well-known hill-climbing method.

5.2.3 Iterated Local Search

Iterated Local Search (ILS) meta-heuristic is based on an LS approach that employs a single solution throughout the iterative process. It is controlled by the local search, perturbation, and stopping criterion phases. ILS is fast in terms of computation time, so it enhances the quality of local optimum and is useful for high-dimensional problems. The ILS process typically starts with a random initial solution within the given search space and continues until the stopping condition is reached. This method was employed by Cortés-Murcia et al. [10] to solve the EVRP with time windows, partial recharges, and satellite customers. To enhance this framework, a Variable Search Descending algorithm was used as a local search component with a set partitioning model in the post-optimization phase.

5.2.4 Genetic Algorithm

The Genetic Algorithm (GA) is an evolutionary algorithm inspired by the natural reproduction process. The basic principle is to first produce a random population of individual solutions known as chromosomes and then evolve it over a number of iterations known as generations. By combining two chromosomes from the current generation using a crossover operator or altering a chromosome using a mutation operator, new chromosomes are created to establish the next generation. A new generation is created by choosing certain parents and children based on their fitness values and rejecting others to maintain a constant population size. After multiple generations, the algorithm converges on the best chromosome, which reflects the best solution to the problem.

This popular algorithm was hired by many researchers and it was also combined with other approaches to constitute a hybrid method.

5.3 Hybrid Methods

A hybrid algorithm offers the choice between at least two different approaches that might efficiently solve the same problem. A lot of EVRP research papers adopt hybrid algorithms, which combine exact, heuristic, and meta-heuristic methods.

This hybridization may incorporate modules, algorithmic principles, or meta-heuristic notions. Hybrid methods can take one of two structural forms [16]. It can be a straightforward parallel of two algorithms, such as using algorithm A first and B later, or a sequential invocation of the two algorithms, or it can be a full hybrid of the two algorithms into a new algorithm that is inseparable. Shao et al. [24] proposed an EVRP model based on a hybrid genetic algorithm to obtain the routes and charging plan mixed with a dynamic Dijkstra algorithm to locate the most energy-efficient paths between any two neighboring visited nodes in the routes.

Keskin and çatay [25] formulated the EVRP with Time Windows and Fast Chargers (EVRPTW-FC) as a mixed integer linear program and solved the small instances using CPLEX, while for the large instances, the researchers developed a two-phase meta-heuristic approach which matches the Adaptive Large Neighborhood Search (ALNS) with an exact method.

The same for Zhou et al. [26] who has used a mixed integer linear programming for small-scale scenarios, while for large-scale instances faced by real-world applications, a hybrid meta-heuristic was developed via integrating a modified Greedy Algorithm with a Variable Neighborhood Search (VNS).

Erdoğdu and Karabulut [27] studied the Electric Traveling Salesman Problem with Time Windows where a Simulated Annealing algorithm was hybridized with a local search heuristic and another constructive heuristic.

The two-echelon EVRP was attractive for some researchers. Breunig et al. [11] have introduced this variant and proposed a Large Neighborhood Search (LNS) meta-heuristic as well as an exact mathematical programming algorithm.

5.4 Review Summary

After an exhaustive literature review, it becomes apparent that multiple approaches have been proposed to address the multi-objective EVRP, each with its own unique characteristics. The range of techniques available is diverse, catering to the specific requirements and complexities of the problem.

Table 1 gives an overview of the different solution methods considered in the reviewed papers, regardless of the problem variant or the constraints taken into

Table 1 Exploration of the MO-EVRP: related studies and research

Problem variant	Objective functions	Charging strategy	Solution method	Algorithm	Reference
Two-echelon EVRP	1. Minimize the traveled distance 2. Minimize the number of EVs	Full charging	Exact method	Branch-and-price algorithm	[21]
EVRP with nonlinear charging time	1. Minimize the travel time 2. Minimize the charging times	Full charging	Exact method	Branch-and-price algorithm	[20]
EVRP with time of use energy pricing and partial recharging	1. Minimize travel distance 2. Minimize the charging cost	Partial charging	Exact method	Mixed integer programming	[22]
EVRP with time windows and stochastic waiting time at charging stations	1. Minimize the number of EVs 2. Minimize the energy cost 3. Minimize the driver wage	Partial recharging	Meta-heuristic method	Adaptive large neighborhood search	[9]
EVRP with time windows	1. Minimize the number of EVs 2. Minimize the traveled distance	Full charging	Meta-heuristic method	Adaptive large neighborhood search	[23]
EVRP with time windows and satellite customers	1. Minimize the time spent at stations 2. Minimize the travel distance	Partial charging	meta-heuristic method	Iterated local search	[10]
EVRP with time windows and fast chargers	1. Minimize the number of EVs 2. Minimize the charging cost	Partial charging	Hybrid method	Adaptive large neighborhood search with exact method	[25]

Electric traveling salesman problem with time windows	1. Minimize the travel distance	Full charging	Hybrid method	Simulated annealing with constructive heuristic and local search	[27]
	2. Minimize the energy consumption				
Two-echelon EVRP	1. Minimize the travel distance	Full charging	Hybrid method	Large neighborhood search with exact method	[11]
	2. Minimize the vehicle cost				
EVRP with partial recharge and vehicle recycling	1. Minimize the vehicle cost	Partial charging	Hybrid method	Greedy algorithm with variable neighborhood search	[26]
	2. Minimize the travel distance				
	3. Minimize the waiting cost				
	4. Minimize the charging cost				

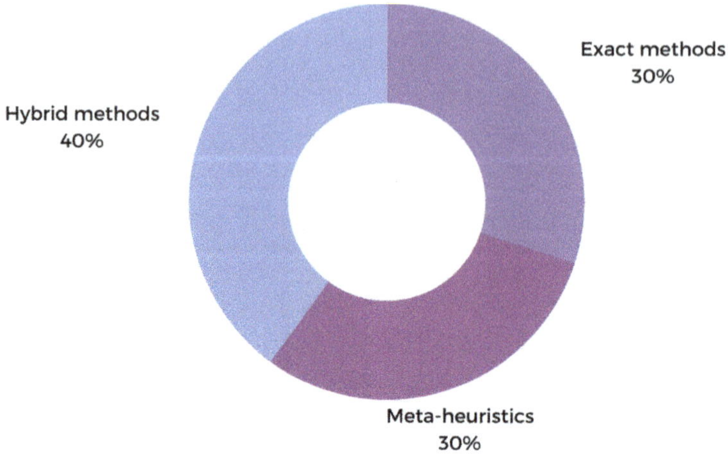

Fig. 4 Distribution of solution method utilization for the MO-EVRP

account. Additionally, Fig. 4 shows the percentage of the usage rate of each solution approach used in the reviewed papers. Among the various approaches, one can find meta-heuristics that combine the strengths of exact methods with the adaptability and exploration capabilities of meta-heuristics. These hybrid algorithms leverage the mathematical guarantees of exact methods while benefiting from the flexibility and efficiency of meta-heuristics. Additionally, sim-heuristics have emerged as a promising approach that combines simulation techniques with meta-heuristics.

By integrating simulation models into the optimization process, sim-heuristics provide a more realistic representation of the problem, enabling more accurate evaluations and improved solution quality. Lastly, the advent of learn-heuristics has introduced a fusion of machine-learning techniques with meta-heuristics. These innovative approaches leverage machine-learning algorithms to enhance the decision-making capabilities of meta-heuristics, allowing them to learn from past experiences and adapt their search strategies dynamically. As far as it can be told, it is recognized that not any research paper has suggested a learning heuristic for solving the EVRP. Since the EVRP is a kind of variant of the VRP, the state-of-the-art of some learn-heuristics for solving this latter problem, which is rare and represents also a minority, is reported. Pugliese et al. [28] have developed an approach inspired by Variable Neighborhood Search (VNS), where several machine-learning techniques are used to deal with the VRP with crowd-shipping. In the same context, Bayliss [29] has proposed a learn-heuristic that integrates machine learning to predict travel times for urban routing problems. Another learn-heuristic was suggested by Bayliss et al. [30] to solve the team orienteering problem—an attractive variant of VRP—which differs from the fact that not all customers have to be visited.

Overall, the EVRP research landscape encompasses a diverse range of approaches, including hybrid meta-heuristics, sim-heuristics, and learn-heuristics,

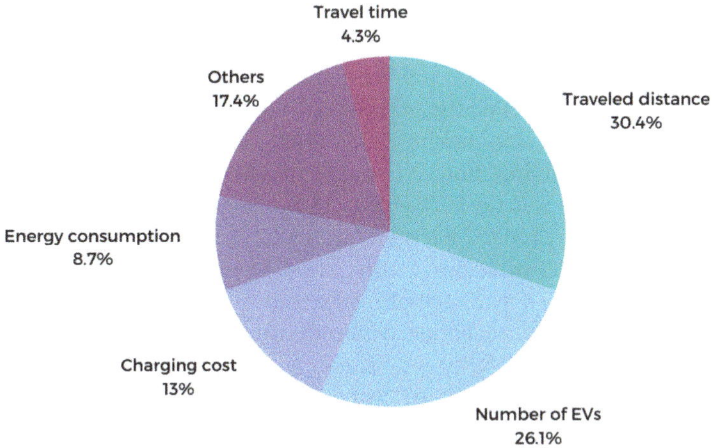

Fig. 5 Frequency of usage for principal objective functions in the MO-EVRP

each contributing unique strengths to the solution process. Continued exploration and advancement of these techniques hold promise for overcoming the challenges posed by the multi-objective EVRP and achieving more efficient and sustainable electric vehicle routing solutions.

Apart from these approaches, several objective functions can be distinguished as shown in Fig. 5. Most papers that are interested in working on the MO-EVRP, consider the travel distance as encountered in the traditional VRP. In addition to the distance objective function, some of the papers considered the number of EVs or the vehicle cost. Other frequently used constraints for the MO-EVRP are the charging cost, the energy consumption, the travel time, and some other operational costs.

6 Challenges and Future Research Directions

One of the primary challenges in the EVRP is the limited availability of accurate and reliable data pertaining to electric vehicle characteristics, traffic conditions, and charging station availability. To address this challenge, future research should emphasize the development of data-driven models and algorithms capable of handling uncertainties and variations in the input data. These models should incorporate techniques such as machine learning, data analytics, and stochastic optimization to improve the accuracy and reliability of the EVRP solutions. Another crucial challenge is the integration of EVRP with other transportation and logistics problems within the broader supply chain context. By incorporating EVRP into a comprehensive framework that considers vehicle routing with time windows, inventory management, and facility location, researchers can optimize the overall supply chain performance. This integration presents an opportunity to develop innovative optimization models and algorithms that can effectively address the complexities and interdependencies of multiple interconnected problems. Furthermore, the advent of intelligent and autonomous vehicles holds great potential for the future of EVRP. Intelligent vehicles equipped with advanced sensing, communication, and decision-making capabilities can enable real-time optimization and dynamic route planning. Future research should explore the use of intelligent and autonomous vehicles in the EVRP context, investigating how these technologies can enhance efficiency, reduce energy consumption, and improve service quality through adaptive routing strategies. Lastly, to validate the effectiveness of proposed models and algorithms, it is essential to test and evaluate them using real-world data. Researchers should collaborate with industry partners and utilize large-scale datasets to assess the scalability, efficiency, and effectiveness of the developed approaches in solving practical EVRP instances.

In summary, future research in EVRP should focus on addressing challenges related to data availability, integration with other logistics problems, leveraging intelligent and autonomous vehicles, developing multi-objective optimization models, and conducting thorough testing and validation. By tackling these areas, researchers can contribute to the advancement of EVRP and facilitate the widespread adoption of electric vehicles in transportation and logistics domains.

7 Conclusion

In this chapter, the concept of multi-objective optimization was introduced, and the main ideas of the EVRP were also examined. A comprehensive literature review of the Multi-objective Electric Vehicle Routing Problem (MO-EVRP) has been presented. Furthermore, various variants of the problem, solution algorithms, and applications of EVRP in different domains such as transportation and logistics were reviewed. The review showed that significant progress has been made in recent years in addressing EVRP and its variants. Some effective models and algorithms have been proposed to solve different variants of the problem. However, there are still many open research questions and opportunities for further improvements and applications of EVRP in real-world settings.

References

1. H. Afshari, W. Hare, S. Tesfamariam, "Constrained multi-objective optimization algorithms: review and comparison with application in reinforced concrete structures. Appl. Soft Comput. J. **83**, 105631 (2019)
2. N. Gunantara, A review of multi-objective optimization: Methods and its applications. Cogent Eng. **5**, 1502242 (2018)
3. H. Liu, Y. Li, Z. Duan, C. Chen, A review on multi-objective optimization framework in wind energy forecasting techniques and applications. Energy Convers. Manag. **224**, 113324 (2020)

4. A.S. Sidhu, S. Singh, R. Kumar, Bibliometric analysis of entropy weights method for multi-objective optimization in machining operations. Mat. Today Proc. **50**, 1248–1255 (2022)
5. M. Caramia, P. Dell'Olmo, Multi-objective optimization, in *Multi-Objective Management in Freight Logistics* (Springer, Cham, 2020), pp. 21–51
6. O. Bahri, E.-G. Talbi, L. Amodeo, Use of electric vehicles in home-health care routing problems: analysis of a multi-objective approach under uncertainty. IFAC-PapersOnLine **54**, 127–132 (2021)
7. I. Kucukoglu, R. Dewil, D. Cattrysse, The electric vehicle routing problem and its variations: a literature review. Comput. Ind. Eng. **161**, 107650 (2021)
8. G.B. Dantzig, J.H. Ramser, The truck dispatching problem. Manag. Sci. **6**, 80–91 (1959)
9. M. Keskin, B. Çatay, G. Laporte, A simulation-based heuristic for the electric vehicle routing problem with time windows and stochastic waiting times at recharging stations. Comput. Oper. Res. **125**, 105060 (2021)
10. D.L. Cortés-Murcia, C. Prodhon, H. Murat Afsar, The electric vehicle routing problem with time windows, partial recharges and satellite customers. Transport. Res. Part E **130**, 184–206 (2019)
11. U. Breunig, R. Baldacci, R.F. Hartl, T. Vidal, The electric two-echelon vehicle routing problem. Comput. Oper. Res. **103**, 198–210 (2018)
12. N.A. Kyriakakis, T. Stamadianos, M. Marinaki, Y. Marinakis, The electric vehicle routing problem with drones: an energy minimization approach for aerial deliveries. Cleaner Logist. Supply Chain **4**, 100041 (2022)
13. S. Bahrami, M. Nourinejad, G. Amirjamshidi, M.J. Roorda, The plugin hybrid electric vehicle routing problem: a power management strategy model. Transport. Res. Part C **111**, 318–333 (2020)
14. M.K. Hasan, M. Mahmud, A.K.M. Ahasan Habib, S.M.A. Motakabber, S. Islam, Review of electric vehicle energy storage and management system: standards, issues, and challenges. J. Energy Storage **41**, 102940 (2021)
15. K.T. Chau, Y.S. Wong, C.C. Chan, An overview of energy sources for electric vehicles. Energy Convers. Manag. **40**, 1021–1039 (1999)
16. C. Ye, W. He, H. Chen, Electric vehicle routing models and solution algorithms in logistics distribution: a systematic review. Environ. Sci. Pollut. Res. **29**, 57067–57090 (2022)
17. H. Qin, X. Su, T. Ren, Z. Luo, A review on the electric vehicle routing problems: variants and algorithms. Front. Eng. Manag. **8**, 370–389 (2021)
18. N. Jozefowiez, F. Semet, E.-G. Talbi, Multi-objective vehicle routing problems. Eur. J. Oper. Res. **189**, 293–309 (2008)
19. S. Hulagu, H.B. Celikoglu, A multiple objective formulation of an electric vehicle routing problem for shuttle bus fleet at a university campus, in *6th International Conference on Models and Technologies for Intelligent Transportation Systems* (2019)
20. C. Lee, An exact algorithm for the electric-vehicle routing problem with nonlinear charging time. J. Oper. Res. Soc. **72**, 1461–1485 (2020)
21. Z. Wu, J. Zhang, A branch-and-price algorithm for two-echelon electric vehicle routing problem. Compl. Intell. Syst. **9**, 2475–2490 (2021)
22. G. Ferro, M. Paolucci, M. Robba, An optimization model for electrical vehicles routing with time of use energy pricing and partial recharging. IFAC-PapersOnLine **51**, 212–217 (2018)
23. T. Erdelic, T. Caric, M. Erdelic, L. Tisljaric, Electric vehicle routing problem with single or multiple recharge. Transport. Res. Procedia **40**, 217–224 (2019)
24. S. Shao, W. Guan, J. Bi, Electric vehicle-routing problem with charging demands and energy consumption. IET Intell. Transp. Syst. **12**, 202–212 (2018)
25. M. Keskin, B. çatay, A matheuristic method for the electric vehicle routing problem with time windows and fast chargers. Comput. Oper. Res. **100**, 172–188 (2018)
26. Y. Zhou, J. Huang, J. Shi, R. Wang, K. Huang, The electric vehicle routing problem with partial recharge and vehicle recycling. Compl. Intell. Syst. **7**, 1445–1458 (2021)
27. K. Erdoğdu, K. Karabulut, Distance and energy consumption minimization in electric traveling salesman problem with time windows, in *7th International Conference on Electrical and Electronics Engineering ICEEE* (2020)

28. L.D.P. Pugliese, D. Ferone, P. Festa, F. Guerriero, G. Macrina, Combining variable neighborhood search and machine learning to solve the vehicle routing problem with crowd-shipping. Optim. Lett. **17**, 1981–2003 (2022)
29. C. Bayliss, Machine learning based simulation optimisation for urban routing problems. Appl. Soft Comput. **105**, 107269 (2021)
30. C. Bayliss, A.A. Juan, C.S.M. Currie, J. Panadero, A learnheuristic approach for the team orienteering problem with aerial drone motion constraints. Appl. Soft Comput. **92**, 106280 (2020)

A Decision Support System for Solving the Windy Rural Postman Problem

Takwa Tlili, Marwa Harzi, and Saoussen Krichen

1 Introduction

The Postman Problem is a classic mathematical problem that involves finding the shortest possible route through a graph or a network, such that every edge is traversed at least once. Over the years, several variants of the Postman Problem (PP) have been developed to address different scenarios and requirements. One popular variant is the Rural Postman Problem, which involves finding the shortest route that covers every edge in a graph while minimizing the total distance traveled. Another variant is the multiple traveling salesman problem, which involves finding the shortest route that allows multiple salesmen to visit a set of cities and return to their starting point. These variants and others have practical applications in a variety of fields, such as logistics, transportation, and computer network optimization. Overall, the Postman Problem and its variants remain important topics of study in the field of optimization and graph theory, with ongoing research and development focused on improving algorithms and finding new applications. Recent years have seen significant research into routing problems with profit, which involves selecting customers from a set of potential customers based on their associated profit when served. The well-known Chinese postman problem involves finding a minimum-cost closed walk that traverses each edge at least once in an undirected and connected graph and is solvable in polynomial time. However, if the cost of traversing an edge from i to j is different from the cost of traversing it from j to i, the problem becomes NP-hard and is known as the Windy Postman Problem (WPP). The Windy Rural Postman Problem (WRPP) is a generalization of the WPP where only the edges in a given subset of required edges need to be traversed. Applications of this problem

T. Tlili (✉) · M. Harzi · S. Krichen
LARODEC, Institut Superieur de Gestion de Tunis, Tunis, Tunisia
e-mail: saoussen.krichen@isg.rnu.tn

I. Alharbi et al. (eds.), *Advances in Computational Logistics and Supply Chain Analytics*, Unsupervised and Semi-Supervised Learning,
https://doi.org/10.1007/978-3-031-50036-7_3

arise in trash collection and newspaper delivery, combining arc routing problems with zigzag options and time dependencies. The objective of these problems is to find one or more routes that optimize an objective function of profit and/or cost. Two fields in the literature assign both profits and costs to arcs, known as CPP with profit and arc routing problems with benefit.

The remainder of this chapter is arranged as follows. Section 2 presents a literature review on the windy rural postman problem. The proposed hybrid neighborhood search-based approach is presented in Sect. 3. In Sect. 4, the designed decision support system is presented and an illustrative example is given in Sect. 5. Section 6 is a real-world application of the evoked problem. Finally, Sect. 7 draws a conclusion and points out possible future work.

2 Related Work

The Windy Rural Postman Problem (WRPP) is a combinatorial optimization problem with various real-world applications, such as waste management, postal services, and snow plowing. The objective is to find the shortest route that covers all edges of an undirected graph, where each edge has a weight and a direction, and the edges may have time windows. Numerous WRPP variants have been evoked in the literature such as the Multi-Vehicle Windy Rural Postman Problem (MV-WRPP) which is an extension of WRPP where multiple vehicles are available to service the nodes. In 2018, [17] proposed a heuristic algorithm based on a hybrid of the Tabu Search and Genetic Algorithm (GA) to solve this problem. Not long ago, [1] proposed an improved version of the Ant Colony Optimization (ACO) algorithm to solve MV-WRPP. Another variant that has been handled is the Time-Constrained Windy Rural Postman Problem (TC-WRPP) which imposes time constraints for visiting each node. In 2019, [18] developed a novel heuristic algorithm based on the Greedy Randomized Adaptive Search Procedure (GRASP). In the same year, [14] proposed a hybrid algorithm based on the Ant Colony Optimization (ACO) and Variable Neighborhood Search (VNS) to solve TC-WRPP. Meng et al. [15] formulated the WRPP as a multi-objective version (MO-WRPP) and proposed a multi-objective particle swarm optimization to solve this variant. Besides, [13] proposed a multi-objective optimization model based on the Non-dominated Sorting GA to solve MO-WRPP. Bouchard and Gendreau [4] addressed the WRPP with multiple vehicles and proposed a decomposition approach based on a new formulation of the problem [8]. The combination of three metaheuristics: Simulated Annealing (SA), Tabu Search (TS), and VNS to handle the WRPP with time windows (WRPPTWs). Gholami and Jabalameli [9] and Asadi et al. [2] proposed a hybrid algorithm that combines a GA and a local search method to find near-optimal solutions for the WRPPTW. Atay and Yaman [3] proposed an evolutionary algorithm with crossover, mutation, and local search operators to solve the Windy Rural Postman Problem with time windows. Huang et al. [11] designed a new heuristic based on two phases of path construction and improvement to solve the Windy Rural Postman Problem with time windows. Jalali et al. [12] developed a hybrid heuristic algorithm based on a combination of simulated annealing, tabu search, and genetic algorithm to solve the Windy Rural Postman Problem. More recently, [7] tackled the WRPP using a novelty search approach by exploring and rewarding novel solutions. Another hybrid approach that integrates three metaheuristics for solving efficiently the WRPP was proposed by Cai and Zhou [5]. The authors tested their algorithm on a set of benchmark instances and compared it with other state-of-the-art algorithms. Wu et al. [16] designed an algorithm that combines a constructive heuristic and a perturbation procedure to generate initial solutions and then applies an adaptive large neighborhood search to improve the solutions.

In summary, these studies show that the Windy Rural Postman Problem is a challenging problem with various real-world applications, and different algorithms can be used to solve it effectively, including hybrid heuristics, evolutionary algorithms, reformulation, decomposition approach, and novelty search algorithms.

3 Neighborhood Search-Based Approach

Neighborhood search is a type of local search algorithm used to optimize a given solution by iteratively searching its neighboring solutions. It is widely used in various optimization problems, including scheduling, routing, and logistics. It is about finding good or near-optimal solutions to an optimization problem. The search tries to improve the current solution by looking for a better solution that is in its neighborhood. In that sense, the neighborhood of the current solution includes a possibly large number of solutions that are near the current solution.

Given an instance I of a combinatorial optimization problem where X is the set of feasible solutions for the instance and $c : X \to R$ is a function that maps from a solution to its cost. Let us consider a minimization problem, and we want to find a solution x^* such that $c(x^*) \leqslant c(x) \forall x \in X$. We define a neighborhood of a solution $x \in XasN(x) \subseteq X$. That is, N is a function that maps a solution to a set of solutions. A solution x is said to be a local optimum with respect to a neighborhood N if $c(x) \leq c(x0) \forall x0 \in N(x)$.

A neighborhood search algorithm typically starts with an initial feasible solution and repeatedly replaces it with an improved solution until some termination criterion is satisfied [10]. In the past decade, the operations research domain has witnessed a strong interest in the development of neighborhood search algorithms. Neighborhood search algorithms are viewed as an effective tool to solve NP-hard combinatorial optimization problems [6].

To tackle the WRPP with time windows, we develop a metaheuristic that combines VNS and TS. For each new generation of schedules, the key is to hybridize the local improvement ability of some VNS and TS to balance exploration and exploitation. The framework of the algorithm with the ordering concept is as follows. We retail the proposed approach in the next subsections.

Initialization is a critical phase for any resolution method since the final solution quality is often influenced by the choice of the initial solution. Therefore we propose a new assigning and scheduling vehicle routes algorithm. We compute the minimal number of vehicles (m_0) that could achieve all routes. In this algorithm, we consider the set of m_0 vehicles at the same time.

$$m_0 = \frac{\sum_{k=1}^{K} D(k)}{T_{max}}.$$

Let ϕ be a set of vehicle routes (K) in a decreasing order of their duration. As a first step, the proposed method starts by assigning, one by one, m_0 routes $\in \phi$ to m_0 vehicles. These routes are scheduled at the end of the time window $[e_0, l_{0d}]$ in order to exploit very well the work day of required vehicles. In the second step, the method assigns the first route from the remaining ones to the vehicle that has the largest available time. If the heuristic cannot assign the route to any available vehicles, it engages a new vehicle. This procedure is repeated until there are no more routes.

The initial solution (Algorithm 1) shows that a good balance between the exploitation of vehicle's time windows (increases) and the number of required vehicles (decreases) can be reached by using this approach.

4 Decision Support System

Due to the complexity of planning and transportation problems, there has been a growing interest in using decision support systems (DSSs) to analyze them at the operational, tactical, and strategic planning levels. Adequate graphical interfaces are important to represent solutions to routing problems. Our DSS is based on tool optimization (TS and VNS algorithm) and Google Earth application that satisfies all customer requests trying to optimally generate vehicles' paths and to minimize the total travel distance and the total used vehicles by respecting the double time window constraint. The methodology can be summarized in three interconnected steps. Such steps are stated as follows.

4.1 Data Inputs

As depicted in Fig. 1, the DSS starts by inputting general problem parameters, namely the number of customers to be served and the number of available vehicles. Once these data are provided, geographical and time coordinates are to be set. Customers' geographical coordinates are extracted from the original distance matrix. Time window coordinates contain basically the earliest time for traveling (e_0), the latest time for traveling (l_{0t}), and the latest time for distribution (l_{0d}).

Data:
$T_{max} \leftarrow$ driver maximum working day hours
$e_0 \leftarrow$ the earliest time for travelling
$l_{0t} \leftarrow$ the lastest time for travelling
$l_{0d} \leftarrow$ the lastest time for distribution
begin
STEP 1
Construct a set ϕ of vehicle routes in decreasing order of their duration
STEP 2
Compute the minimal number of vehicles m_0
STEP 3
(1) Assign, one by one, routes $\in \phi$ to m_0 vehicles.
(2) Compute for each route $T(k)$ assigned to vehicle m its start date:
$DD_m(k) = l_{0t} - max(0, (D(k) - (l_{0d} - l_{0t})))$.
(3) Schedule routes at the end of the time window $[e_0, l_{0r}]$.
(4) Compute for each vehicle m assigned its available time:
$\Delta T(m) = min(T_{max} - D(k), DD_m(k) - e_0)$.
STEP 4
Assign the first route from the remaining ones ($T(k) \in \phi$) to vehicle m (among m_0) that has the largest available time $\Delta T(m)$.
if $\Delta T(m) \geq D(k)$ **then**
Assign $T(k)$ to vehicle m:
$DD_m(k) = \Delta T(m) - D(k)$ and $\Delta T(m) = DD_m(k)$;
else
$m_0 = m_0 + 1$;
Compute $DD_m(k)$ and $\Delta T(m)$ according to equations (6) and (7). Assign $T(k)$ to $m = m_0$ and schedule route at the end of time window $[e_0, l_{0d}]$.
end
STEP 5
Let s be the scheduled and the assigned routes $\in T$ to vehicle m, and m_0 the number of required vehicles.
end

Algorithm 1: Assigning and scheduling vehicle routes algorithm (ASVRA)

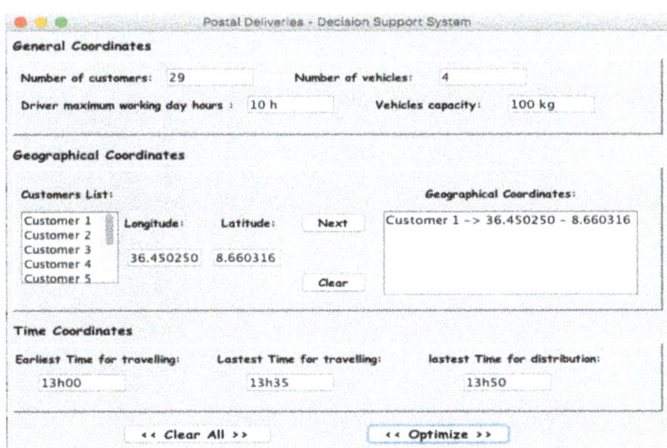

Fig. 1 Data Inputs: general, geographical, and time coordinates

4.2 Optimization Tools

This module is the kernel of the DSS due to its significant contribution to decision-making. The optimization approach is modeled to solve the defined VRPDTW variant by an exact method or an approximate algorithm. The decision-maker can choose between the optimal solution proposed by the CPLEX solver and the near-optimal solution given by the TS-VNS (in our case, we have just encoded the approximate method). Further explanations for each proposed solution are offered by the system to assist the DM in taking the best alternative.

4.3 Numeric and Cartographic Display Solution

As shown in Fig. 2, once the numerical solution is generated, the DSS moves to the design of the cartographical solution that well illustrates the real itinerary. Customers' locations are then marked in the addressed area and vehicles' pathways are highlighted.

5 Illustrative Example

The next example illustrates the process of the proposed algorithm. We consider the following data:

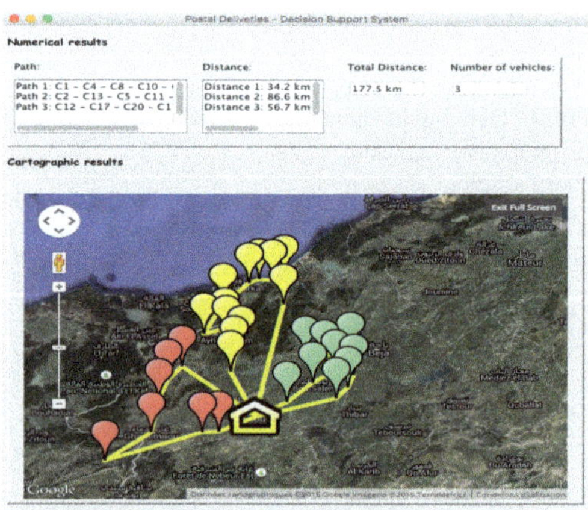

Fig. 2 Optimization results: numerical and geographical solution

Table 1 Duration carried on vehicle routes

Routes	$T(1)$	$T(2)$	$T(3)$	$T(4)$	$T(5)$
$D(k)(h)$	6	4	3	2	1

The time parameters are defined as follows: $T_{\max} = 10h$, $e_0 = 8h$, $l_{0t} = 15h$, $l_{0d} = 18h$, while the remaining parameters are listed in Table 1.

- **Step 1:** $\phi = \{T(1), T(2), T(3), T(4), T(5)\}$
- **Step 2:** $m_0 = \frac{\sum_{k=1}^{5} D(k)}{T_{max}} = 2$
- **Step 3:**
  1. T(1) and T(2) are assigned to vehicles 1 and 2.
  2. $DD_1(1) = 12\,\text{h}$ and $DD_2(2) = 14\,\text{h}$
  3. Schedule routes (Fig. 3)
  4. $\Delta T(1) = 4\,\text{h}$ and $\Delta T(2) = 6\,\text{h}$

- **Step 4:**
  1. $\phi = \{T(3), T(4), T(5)\}$, $\Delta T(2) > \Delta T(1)$, and $D(3) < \Delta T(2)$. Therefore $T(3)$ is assigned to vehicle 2 with $DD_2(3) = 11$ and $\Delta T(2) = 3\,\text{h}$ (Fig. 4).
  2. $\phi = \{T(4), T(5)\}$, $\Delta T(1) > \Delta T(2)$, and $D(4) < \Delta T(1)$. Consequently $T(4)$ is assigned to vehicle 1 with $DD_1(4) = 10$, and $\Delta T(1) = 2\,\text{h}$ (Fig. 4).
  3. $\phi = T(5)\}$, $\Delta T(2) > \Delta T(1)$ and $D(5) < \Delta T(2)$. Then $T(5)$ is assigned to vehicle 2 with $DD_2(5) = 10$ and $\Delta T(2) = 2\,\text{h}$ (Fig. 4).

- **Step 5:** The solution s (Figs. 3 and 4) represents the initial solution (the assigned routes $\in T$ to vehicles m using m_0 vehicles) given by Algorithm 1.

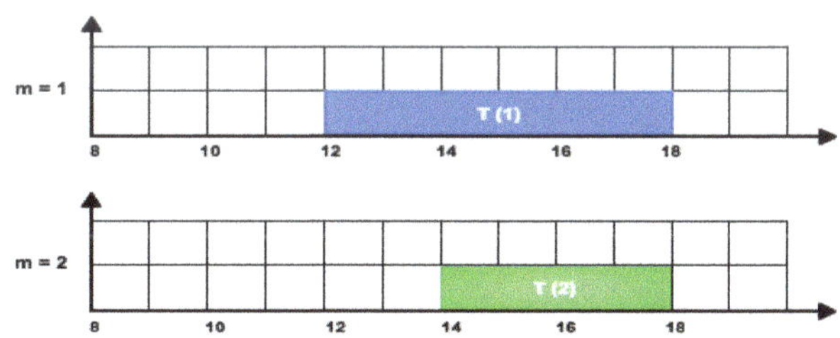

Fig. 3 Assigning process of $T(1)$ and $T(2)$ to vehicles 1 and 2, respectively

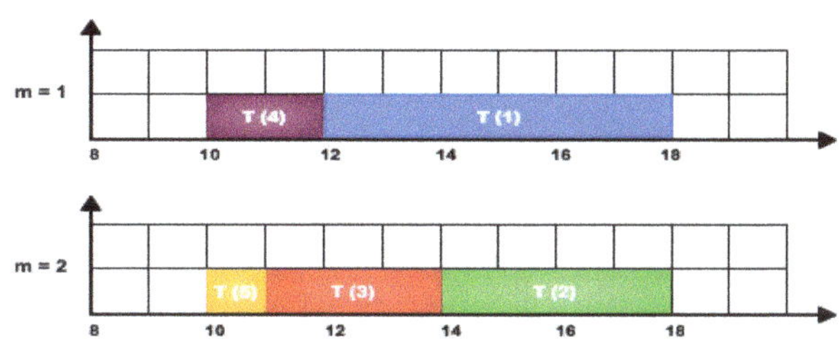

Fig. 4 Assigning process of $T(3)$, $T(4)$, and $T(5)$ to vehicles 2, 1, and 2, respectively

5.1 Neighborhood Search

Since the complementary neighborhood functions are the critical idea behind VNS, the neighborhood structure (NS) should be chosen very rigorously to achieve an efficient VNS. Here, we tried to propose several NSs with different designs for generating diverse solutions in the fastest and easiest way possible. The proposed neighborhood structures (Permute, Swap, and Insert Operator) are listed below. The following example illustrates the presented movements and neighborhood search. Let C_i be the customers belonging to routes $T(k)$ (Tables 2 and 3), represents the solution for the vehicle route configurations of Table 2 using ASVRA algorithm (Algorithm 1).

1. Permute Operator: This operator consists in swapping two randomly selected routes. This operator plays a key role because it influences not only route configurations but also scheduled vehicle routes. To illustrate this, let us permute routes 4 and 5 in the set of routes of Table 2 as shown in Fig. 5. After applying the ASVRA algorithm to Fig. 5, we obtain the solution in Table 4. It is interesting to note that one vehicle has been saved in this illustration.
2. Swap Sequence Operator: The swap operator permutes a number of customers in the route configurations. First, two cut points are randomly selected on the route

Table 2 Vehicle route configurations ($T_{max} = 10$)

	C_i	$D(k)(Hours)$
$T(1)$	$0-6-9-10-12-19-0$	5.5
$T(2)$	$0-2-7-1-0$	3
$T(3)$	$0-16-4-0$	5
$T(4)$	$0-3-13-11-0$	2
$T(5)$	$0-15-5-8-0$	4.5
$T(6)$	$0-9-14-0$	3.5
$T(7)$	$0-17-18-20-0$	4.5

Table 3 Solution for the vehicle route configurations of Table 2 using ASVRA algorithm (Algorithm 1)

Vehicles	Routes	$D(k)(Hours)$
1	1	5.5
	2	3
2	3	5
	4	2
3	5	4.5
	6	3.5
4	7	4.5

Route 1				Route 2			Route 3		Route 4			Route 5			Route 6		Route 7		
6	10	12	19	2	7	1	16	4	15	5	8	3	13	11	9	14	17	18	20

Fig. 5 Randomly permute two routes illustration

Table 4 Solution for the vehicle route configurations of Fig. 5 using ASVRA algorithm

Vehicles	Routes	$D(k)(Hours)$
1	1	5.5
	2	3
2	3	5
	4	2
3	7	4.5
	6	3.5
	5	2

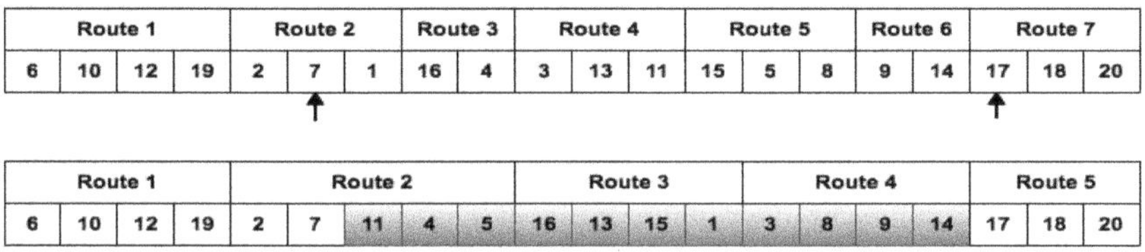

Fig. 6 Swap sequence illustration

configurations. Then, customers located between the two cut points are randomly swapped. Figure 6 shows the route configurations and its result using the swap sequence operator. Cut points are located on customers C_7 and C_{17}. Note that in this illustration the swap sequence operator helped to the minimization of the number of routes.

5.2 *Tabu Search Details*

- **Tabu list**: The tabu list maintains the latest moves of p iterations and a move cannot be done if it is in the tabu list. Actually, a tabu list is used to prevent obtaining cycling solutions. In our algorithm, if moves are not leading to improvement, it is stored in the tabu list in order to prevent returning to already visited solutions in a certain number of algorithm iterations.
- **Tabu tenure** Theoretically, the tabu list needs to store all previously visited solutions. The length of the tabu list can be fixed or variable. The literature showed that the results of variable tabu lists were not significantly better than the fixed ones. To determine the length of the tabu list, we performed some tests with a small number of problem instances. Based on the obtained results, the value of tabu tenure is fixed during the entire search and set to 12.
- **Aspiration criterion** A tabu move becomes admissible if it yields a solution that is better than an aspiration value. In this chapter, we set the aspiration level by the best solution found so far. Therefore, if any move leads to a solution better than the aspiration level, disregarding it being tabu or not, the move will be made and the aspiration level will be updated to the new value.

6 Real-World Application

The Tunisian Post, the trade name of the national office of the Tunisian Posts, is the public company responsible for the postal service in Tunisia. It is an industrial and commercial institution centered on two main activities: (1) the collection and (2) the transport and distribution of mail. Additionally, it offers financial services along with traditional offerings like stamp production and sales, as well as newer services including various computer-related services.

In our real case, we apply the proposed DSS for the postal organization in the region of Jendouba in the northwest of Tunisia (Fig. 7) that needs to restructure its package and mail delivery systems (its core services) in efforts to reduce the delivery distance and a number of used vehicles.

As depicted in Fig. 7, the postal sorting center (the area surrounded by the small green circle) is where the post mail (postal letters, postal packages, etc.) will be collected. The area surrounded by the small black circle (the postal organization of Jendouba) receives postal mail from the postal sorting center, which will be subsequently distributed to the post offices that are in the geographical area encircled by the large red circle (which is the geographical area of the practical case study that will be treated in this chapter).

Let us consider an example of $n = 29$ customers, four vehicles, and one depot dispersed around the Jendouba city as shown in Fig. 2. Different setting parameters of the problem are presented in Table 5 and temporal coordinates are in Table 6. To pick out the shortest path between each couple of customers, we listed in Table 7 the geographical coordinates. Using Table 7, we obtain the matrix that summarizes all shortest paths between each pair of customers and depots.

Based on Table 8, we reported the total distance (kilometer) and the total cost (Tunisian dinar) of each itinerary. We can note from this table that the improvement

Fig. 7 The treated geographical area: the northwest of Tunisia

Table 5 Description of example parameters

Parameters	
Vehicle capacity (kg)	100
Driver maximum working day hours	10
Maximum distance traveled by each vehicle (km)	1000

Table 6 Temporal requirements

i	e_0	l_{0t}	l_{0d}	S_i
Sidi Meskine	08 : 15	08 : 40	08 : 55	15 : 00
Hkim	08 : 55	09 : 30	09 : 50	15 : 00
Oued Mliz	09 : 50	10 : 20	10 : 40	15 : 00
Ghardimaou	10 : 40	11 : 10	11 : 30	15 : 00
Ouerguech	11 : 30	11 : 55	12 : 05	15 : 00
Oued El Maaden	12 : 05	12 : 30	12 : 45	15 : 00
Eddkhailia	12 : 45	13 : 15	13 : 30	15 : 00
Fernena	08 : 15	08 : 35	08 : 45	15 : 00
Jantoura	08 : 45	09 : 05	09 : 20	15 : 00
Beni Mtir	09 : 20	09 : 50	10 : 05	15 : 00
Ain Draham	10 : 05	10 : 25	10 : 40	15 : 00
Babouch	10 : 40	11 : 00	11 : 15	15 : 00
Hamam Bourgiba	11 : 15	11 : 35	11 : 50	15 : 00
Jab' Allah	11 : 50	12 : 15	12 : 30	15 : 00
Tabarka	12 : 30	12 : 55	13 : 10	15 : 00
Ain Essobh	13 : 10	13 : 30	13 : 35	15 : 00
Airport	13 : 35	13 : 55	14 : 10	15 : 00
Hay El Morjen	14 : 10	14 : 25	14 : 40	15 : 00
Brirem	14 : 40	15 : 00	15 : 15	15 : 00
Bouselem	08 : 15	08 : 40	08 : 50	15 : 00
Hay Eroumeni	08 : 50	09 : 15	09 : 30	15 : 00
Bouaouene	09 : 30	09 : 50	10 : 10	15 : 00
Balta	10 : 10	10 : 35	10 : 50	15 : 00
Esomran	10 : 50	10 : 15	10 : 30	15 : 00
Badrouna	10 : 30	10 : 50	11 : 00	15 : 00
Souk Essebet	11 : 00	11 : 20	11 : 35	15 : 00
Ben Bachir	11 : 35	11 : 55	12 : 10	15 : 00
Souk Ejemaa	11 : 10	12 : 35	12 : 55	15 : 00
Jendouba	12 : 55	13 : 15	13 : 35	15 : 00

of the solution amounts to 40.93% for the distance traveled by the vehicle, 14.66% for the costs, and one vehicle has been saved.

For more clarification on the improvement of the obtained solutions, we tested the DSS for five different days by comparing the actual route schedules obtained by the DSS and the routes that were scheduled. The resulting graphs can be seen in Figs. 8 and 9.

Figure 10 displays a geographical view of the obtained results after solving the example. It illustrates the best traveling path for each vehicle while taking into account the time windows, the capacity restriction, and the limited number of vehicles assigned to each depot. This map is used to guide vehicle drivers to serve customers through the shortest itinerary presented as yellow arrows.

In this example, three vehicles are used. The first one leaves the depot to serve in the order Hkim, Sidi Meskine, Oued Mliz, Ghardimaou, Oued el Maaden,

Table 7 Geographical coordinates for the Tunisian case study

i	Longitude	Latitude
Sidi Meskine	36.450250	8.660316
Hkim	36.459729	8.611703
Oued Mliz	36.454206	8.549561
Ghardimaou	36.446712	8.437591
Ouerguech	36.476702	8.449582
Oued El Maaden	36.366722	8.433572
Eddkhailia	36.354321	8.418741
Fernena	36.650379	8.699348
Jantoura	36.711572	8.716354
Beni Mtir	36.742151	8.736105
Ain Draham	36.774642	8.687074
Babouch	36.799335	8.657994
Hamam Bourgiba	36.768777	8.580869
Jab' Allah	36.873624	8.729061
Tabarka	36.943950	8.754301
Ain Essobh	36.954487	8.868618
Airport	36.980512	8.878927
Hay El Morjen	36.955584	8.840465
Brirem	36.980820	8.881321
Bouselem	36.610790	8.974724
Hay Eroumeni	36.594371	8.956235
Bouaouene	36.693714	9.008017
Balta	36.573970	8.919156
Esomran	36.556322	8.881390
Badrouna	36.535912	8.856671
Souk Essebet	36.501146	8.834699
Ben Bachir	36.519911	8.814099
Souk Ejemaa	36.511633	8.801740
Jendouba	36.500552	8.777677

Eddkhailia, and then Ouerguech with a cost of 99.145. The second vehicle serves, respectively, Fernena, Beni Mtir, Jantoura, Ain Draham, Hamam Bourgiba, Babouch, Jab' Allah, Tabarka, Ain Essobh, Hay El Morjen, Airport, Brirem, Souk Essebet, Ben Bachir, and Souk Ejemaa with a cost of 251.053. And the third one leaves the depot to serve Bouselem, Bouaouene, Hay Eroumeni, Balta, Esomran, and Badrouna for 164.373. All of these vehicles come back to the depot Jendouba after serving the corresponding customers. All the results show that using our proposed DSS the company can save from 27.47% to 31.5% in the distance traveled by vehicles and from 31.65% to $35,5$ in distance costs.

Table 8 Example of solution improvement

Vehicle	Itinerary	Distance (km)	Cost (DT)
DSS solution			
1	Jendouba → Sidi Meskine → Hkim → Oued Mliz → Ghardimaou → Ouerguech → Oued el Maaden → Eddkhailia → Jendouba	34.2	99.145
2	Jendouba → Fernena →Jantoura → Beni Mtir → Ain Draham → Babouch →Hamam Bourgiba → Jab' Allah → Tabarka → Hay El Morjen → Ain Essobh → Airport → Brirem → Jendouba	86.6	251.053
3	Jendouba → Bouselem → Hay Eroumeni → Bouaouene → Balta → Esomran → Badrouna → Souk Essebet → Ben Bachir → Souk Ejemaa → Jendouba	56.7	164.373
Current solution			
1	Jendouba → Hkim → Sidi Meskine → Oued Mliz → Ghardimaou → Oued el Maaden → Eddkhailia → Ouerguech → Jendouba	54.5	157.995
2	Jendouba → Fernena → Beni Mtir → Jantoura → Ain Draham → Hamam Bourgiba Babouch → Jab' Allah → Jendouba	40.31	48.58
3	Jendouba → Tabarka → Ain Essobh → Hay El Morjen → Airport → Brirem → Souk Essebet → Ben Bachir → Souk Ejemaa → Jendouba	120.31	148.778
4	Jendouba → Bouselem → Bouaouene → Hay Eroumeni → Balta → Esomran → Badrouna → Jendouba	85.4	247.574

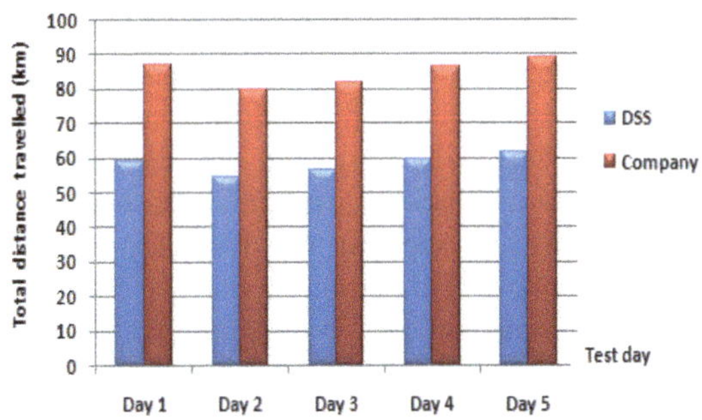

Fig. 8 Comparison between the distance in kilometers traveled by the routes scheduled and the routes obtained by the DSS

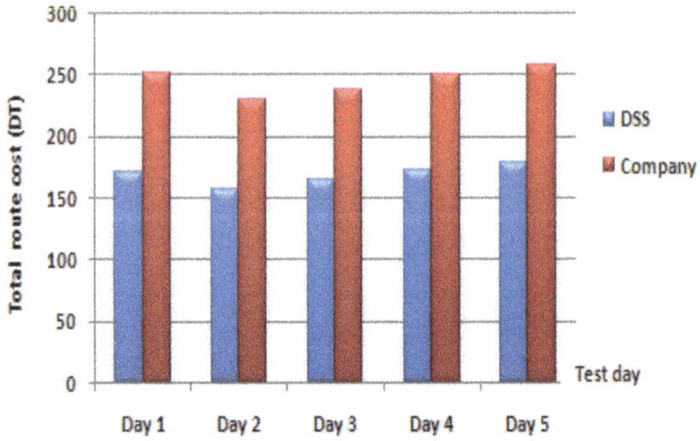

Fig. 9 Comparison between the cost of the routes scheduled by the company and the routes obtained by the DSS

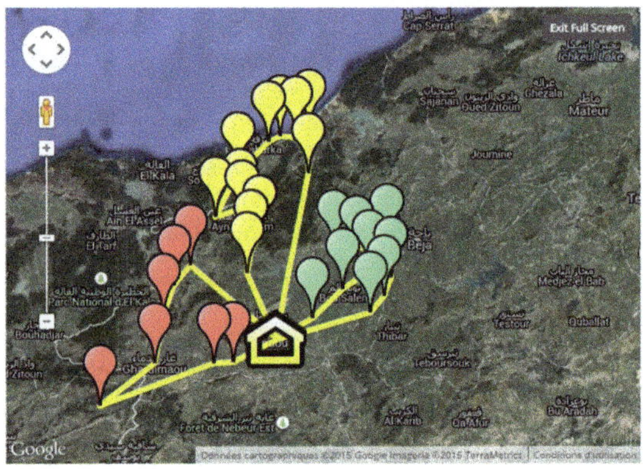

Fig. 10 Geographical solution of the Tunisian case study using the DSS

7 Conclusion

This chapter presented a real-world distribution problem faced by postal companies, which can be modeled as a variant of the windy rural postman problem. Given the NP-hard nature of this problem, approximate methods must be used to find near-optimal solutions within a reasonable amount of time. To address this issue, we proposed a hybrid variable neighborhood-based approach and demonstrated its efficiency by applying it to real-life instances of a postal company in northwest Tunisia. The results show that our algorithm can reduce the number of vehicles required for deliveries while still ensuring timely delivery to customers.

In future works, we will focus on the development of a hybrid variable neighborhood-based approach, showcasing its efficiency and applicability through real-life instances of a postal company in northwest Tunisia. Finally, from a practical and applied standpoint, the chapter underscores the relevance of the proposed solution to the postal industry, highlighting its potential to reduce vehicle requirements while ensuring timely delivery and its potential for implementation in real-world scenarios.

References

1. M. Ahmadizadeh, M. Zamanipour, S.H. Zegordi, An improved ant colony optimization for the multi-vehicle windy rural postman problem. J. Ind. Syst. Eng. **13**(3), 44–61 (2020)
2. S. Asadi, F. Jolai, M. Zandieh, A new hybrid algorithm for solving the windy rural postman problem. J. Ambient Intell. Hum. Comput. **12**(6), 5797–5810 (2021)
3. A.T. Atay, H. Yaman, An evolutionary algorithm for the windy rural postman problem with time windows. Appl. Soft. Comput. **93**, 106311 (2020)
4. R. Bouchard, M. Gendreau, Windy rural postman problem with multiple vehicles. Transp. Res. Part C Emerging Technol. **99**, 75–93 (2019)
5. Y. Cai, M. Zhou, A hybrid heuristic algorithm for the windy rural postman problem. J. Comput. Sci. **31**, 93–104 (2019)
6. V.-D. Cao, M. Gendreau, T.-D. Nguyen, A novel hybridization of large neighborhood search and iterated local search for the multi-depot vehicle routing problem with time windows. IEEE Trans. Evol. Comput. **25**(5), 901–915 (2021)
7. E. Eiben, L. Kovács, Novelty search algorithm for the windy rural postman problem. J. Heuristics **27**(3), 451–471 (2021)
8. T.A. Feo, M.G. Resende, Hybrid heuristic for the windy rural postman problem with time windows. Comput. Oper. Res. **108**, 216–230 (2019)
9. M. Gholami, M.S. Jabalameli, A hybrid algorithm for solving the windy rural postman problem with time windows. Neural Comput. Appl. **31**(11), 7543–7555 (2019)
10. M.A. Gomez-Corona, A.G. Aguirre-Rodriguez, J.J. Rangel-Magdaleno, A variable neighborhood search algorithm for the periodic vehicle routing problem. Comput. Ind. Eng. **165**, 107395 (2022)
11. H. Huang, J. Chen, Q. Zhu, A new heuristic algorithm for the windy rural postman problem with time windows. Math. Probl. Eng. **2019**, 1–9 (2019)
12. M. Jalali, B. Amiri, N. Safaei, A hybrid heuristic algorithm for the windy rural postman problem. J. Appl. Res. Technol. **17**, 162–171 (2019)
13. H. Kaur, H. Kundra, R. Arora, Multi-objective optimization for the windy rural postman problem using NSGA-II. Int. J. Adv. Intell. Paradigms **15**(1), 46–67 (2020)
14. S. Li, W. Huang, H. Yang, A hybrid ant colony optimization and variable neighborhood search algorithm for the time-constrained windy rural postman problem. J. Intell. Manuf. **30**(1), 337–350 (2019)
15. X. Meng, H. Li, X. Li, B. Li, Y. Feng, A multi-objective particle swarm optimization for the multi-vehicle windy rural postman problem. IEEE Access **8**, 124201–124213 (2020)
16. J. Wu, J. Chen, Y. Tan, A hybrid algorithm based on adaptive large neighborhood search for the windy rural postman problem with time windows. J. Ambient. Intell. Humaniz. Comput. **12**(6), 5899–5912 (2021)
17. M. Zamanipour, M. Azimi, S.H. Zegordi, A hybrid tabu search and genetic algorithm for the multi-vehicle windy rural postman problem. J. Ind. Syst. Eng. **11**(4), 79–95 (2018)
18. Z. Zhou, J. Hu, Y. Liu, A grasp algorithm for the time-constrained windy rural postman problem. J. Intell. Manuf. **30**(5), 1863–1873 (2019)

A Hybrid Meta-Heuristic to Solve Flexible Job Shop Scheduling Problem

Makram Zaidi, Amina Amirat, Bassem Jarboui, and Abdelkrim Yahyaoui

1 Introduction

In a lot of industrial systems, scheduling problem is highly crucial [14], and for this reason, this subject has been explored for the last few decades [3–5, 10, 17, 29]. Job shop scheduling problem (JSP) encompasses a subdivision of production scheduling as well as combinatorial optimization problems [15]. Accordingly, the flexible job shop scheduling problem (FJSP) is an expanded version of the job shop scheduling problem (JSP) [16]. FJSP differs from JSP in the sense that it allows the processing of operations on more than one candidate machine, resulting in two sub-problems. These problems are machine assignment and operation sequencing. Specifically, machine assignment relates to the assignment of a machine per operation, whereas operation sequencing relates to the scheduling of every operation on machines for the optimization of the indicators of performance in question [16]. Hence, as opposed to the traditional JSP, FJSP appears to be more complex, aside from being strongly NP hard in 1993 [50].

The common JSP problem with the criterion of makespan is describable using a group of n jobs to be processed on m machines. In each job, there are a number

M. Zaidi
College of Applied Studies and Community Service, Imam Abdulrahman Bin Faisal University, Dammam, Saudi Arabia
e-mail: mmzaidi@iau.edu.sa

A. Amirat (✉) · A. Yahyaoui
College of Business, University of Jeddah, Jeddah, Saudi Arabia
e-mail: aamirat@uj.edu.sa; ayahiaoui@uj.edu.sa

B. Jarboui
College of Economics and Management, University of Sfax, Sfax, Tunisia
e-mail: Jarboui@fsegs.rnu.tn

I. Alharbi et al. (eds.), *Advances in Computational Logistics and Supply Chain Analytics*, Unsupervised and Semi-Supervised Learning,
https://doi.org/10.1007/978-3-031-50036-7_4

of operations, and for a given job, its operations are to be processed according to a specified order. For the operations, machines are employed for an established duration. At one given time, the machines have the capacity to process at maximum one operation at a given time. Here, once the processing of an operation on a particular machine begins, it must proceed smoothly until the end. In this process, the aim is to search for the optimal schedule of the operations on the machines. In this regard, the precedence constraints need to be considered, and this curtails the makespan, whereby the ending time of the preceding operation finishes within the schedule [49].

Brucker and Schlie were the first to have explored FJSP in their application of a polynomial approach in managing two jobs [5]. In handling FJSP, a lot of heuristics or meta-heuristics have been used, particularly via ant colony optimization (ACO) [38], artificial bee colony (ABC) [37], genetic algorithm (GA) [12, 46], particle swarm optimization (PSO) [13, 18], simulated annealing (SA) [2], tabu search (TS) [38], as well as hybrid approaches following differing heuristics and meta-heuristics.

Recent years have seen the growing interest of scholars in FJSP with makespan-oriented performance measures. In this regard, there have been several studies that explore the precise algorithm for resolving FJSP, and among those who first pioneered the resolution of FJSP are Brucker and Schlie (1990) [5]. In their study, Brucker and Schlie constructed a polynomial algorithm for the purpose of resolving this problem using two jobs. In fact, countless meta-heuristics have been developed to solve FJSP. For instance, a tabu search algorithm comprising diverse neighborhood functions was proposed by several researchers (e.g., [4, 10, 37]).

There are also past studies that employed genetic algorithms to FJSP. In this regard, some studies [2, 35, 38, 50] proposed the application of an artificial immune algorithm (AIA) following an integrated approach. In the application of this algorithm, the initial population was fashioned, and individuals for reproduction were chosen. For these purposes, a number of strategies were used. Also, in the reproduction of new individuals, diverse operators of mutation were employed.

In a study conducted by Yazdani et al. [46], the application of a parallel variable neighborhood search (PVNS) algorithm was demonstrated. In this study, the authors attempted to minimize makespan time, and the use of the PVNS algorithm resolved FJSP. The application of PVNS involves the application of diverse neighborhood structures, and these structures transform the operation assignment and sequencing to produce neighboring solutions.

Also, for resolving FJSP, the application of hybrid meta-heuristics has been demonstrated in several studies to minimize the criterion of makespan. Relatedly, the application of a mathematical model and two meta-heuristics algorithms (SA and TS) was demonstrated in [12] for FJSP resolution. In other relevant works, [18] and [13] examined FJSP in the context of three objectives, namely min makespan, min maximal machine workload, and min total workload. In particular, in [19], a Pareto approach was proposed. This approach follows the hybridization of fuzzy logic (FL) alongside evolutionary algorithms, and it is applied in the resolution of multi-objective FJSP. As for [13], the authors demonstrated the application of a hybrid genetic algorithm (hGA) grounded upon the integrated approach.

A study by [41] presented the application of two-phase meta-heuristics (TPM) following the imperialist competitive algorithm (ICA) and variable neighborhood search (VNS). This study has two phases: In the starting phase, the problem is transformed into FJSP with makespan, total tardiness, and total energy consumption. Then, an ICA is used to resolve the new FJSP. In particular, an ICA employs several new methods in constructing initial empires and performing imperialist competition. In the following phase, new strategies are prepared. These strategies compare solutions. At the same time, the non-dominated set of the starting phase is updated. For the original problem, a VNS is employed. Then, to improve the quality of solution, set member is occasionally used in replacement of the current VNS solution. Using optimization, the threshold of energy consumption is acquired. The performance of TPM is tested by way of exhaustive experiments. The obtained outcomes proved the high competences of TPM in resolving FJSP.

The method to resolve FJSP was presented in [34]. The authors called it the hybrid meta-heuristic-based clustered holonic multi-agent model. In demonstrating this method, the authors first employed a neighborhood-based genetic algorithm (NGA) using an agent of scheduler for a global exploration of the search space. Then, for guiding the research in potential regions of the search space as well as for improving the NGA quality of the concluding population, the authors employed a technique of local search using a group of cluster agents. In this study, the efficiency of the proposed approach was justified using the flexible assortment of the potential portions of the search space. For the purpose, the authors employed the operator of clustering following the process of genetic algorithm, as well as the technique of intensification of the tabu search, which enabled the search to be restarted from a group of elite solutions for the achievement of fresh strong scheduling solutions. Results are displayed by way of four groups of renowned benchmark literature instances. In this study, the authors discovered new upper bounds, and this is an evidence of the approach's effectiveness.

In [36], an expansion of the simple job shop scheduling problem (JSSP) operating mode was proposed. The method proposed by the authors has flexibility, that is, its operations can be run using a number of machines. Also, the method includes the use of lot streaming, where jobs may be split into sub-lots. In this study, in order to address the resultant FJSSP-LS, the authors created an integer programming model. The authors applied the model within a commercial solver called GUROBI. Further, the authors proposed a tabu search (TS) algorithm as a technique of solution. A set of 48 instances tailored for the FJSSP-LS working mode was employed for comparison purpose. The authors found that the TS heuristic supersedes the upper bounds by GUROBI which usually would fail to converge to optimality within one hour. This study assists in the application of flexibility and lot streaming in the context of JSSP.

A study by [33] demonstrates the use of particle swarm optimization (PSO) algorithm for resolving FJSP. The main aim of the study was to minimize the maximum accomplishment time criterion. In this study, tests were performed on the numerous benchmark data obtained from the literature. These data include partial FJSP and total FJSP. From the experimental results, the authors concluded the effectiveness and efficiency of the developed PSO in FJSP resolution. In addition, the authors

Table 1 Classification of research papers according to the proposed algorithm

Single solution–based algorithms	Population-based algorithms	Hybrid algorithms	Other
1-Tabu search (TS) **Brandimarte (1993)**[4] **Hunrik et al. (1994)**[17] **Barnes and Chambers (1995)**[3] **Mastrolilli and Gambardella (2000)**[29] **Lee et al. (2007)**[23] **Romero et al. (2018)**[36] **2-Simulated annealing (SA)** **Fattahi et al. (2007)**[12] **Yazdani et al. (2009)**[46] **3- Variable neighborhood search (VNS)** **Yazdani et al. (2010)**[46]	**Ant colony optimization (ACO)** Rossi and Dini (2002) [37] **Artificial bee colony (ABC)** Wang et al. (2017) [41] **Artificial immune system (AIS)** Bagheri et al. (2010) [2] **Evolutionary algorithms (EAs)** Ho and Tay (2004) [15] Zhang and Gen (2005) [50] Ho et al. (2007) [16] Saad et al. (2008) [38] Pezzella et al. (2008) [35] De Giovanny and Pezzella (2010) [11] Fattahi and Fallahi (2007) [12] Lee et al. (2007) [23] **Particle swarm optimization (PSO)** Liu et al. (2022) [27] Nouiri et al (2018) [33]	Dauzere-Peres and Paulli (1997) [10] Kacem et al. (2002a, 2002b) [18, 19] Xia and Wu (2005) [43] Gao et al. (2008) [13] Fattahi et al. (2007) [12] Ho and Tay (2004) [15] Li et al. (2020, 2022) [24, 25] Zhang et al. (2005, 2008, 2019) [48, 49, 50] Nouri et al. (2018) [34]	**1- Deep learning** Chang et al. (2022) [6] Liu et al. (2022) [27] **2- Q-learning** Naimi et al. (2021) [32]

examined the PSO solving method for forthcoming implementation on embedded systems with the capacity of real time decision-making that is based upon the state of resources and all unforeseen events. In this regard, the authors introduced two multi-agent based approaches, and using different benchmark instances, comparisons were made. Table 1 summarizes the most approaches proposed to solve the flexible job shop problem.

This study explores the flexible job shop scheduling problem (FJSSP) for the purpose of minimizing makespan. FJSP is akin to the problem of conventional job shop. The problem consists of organizing the execution of n jobs on m machines. Each job J_i signifies n_i organized operations. The performance of jth operation of job J_i (noted O_{ij}) needs one machine M_k, and this machine is obtained from a group of obtainable machines μ_{kj}(noted $\mu_{kj} \subset M_k$). Accordingly, P_{ijk} signifies the processing time of the operation O_{ij} when performed on machine M_k. It should be noted that each operation must run smoothly without disturbance (non-preemptive condition), and at any given time, each machine can execute only one operation (resource constraint). As a final point, for any job, the antecedence constraints of the operations can be determined for any pair of operations (sequencing constraint). Indeed, this problem is strongly NP hard [14]. As a solution, this study proposes the application of hybrid meta-heuristic algorithm known as genetic algorithm (GA) and simulated annealing (SA). An integrated approach is used.

In [45], authors develop a new immune multi-agent scheduling system (NIMASS) to solve the FJSP with the objective of minimizing the makespan. They simulate humoral immunity to establish the architecture of NIMASS and the negotiation strategies of NIMASS. The results indicate that NIMASS can effectively improve the quality of solution in a very short time.

The work of [7] presents the development of flexible JSS and a consolidated survey of various techniques that have been employed since 1990 for problem resolution with deep evaluation of publications and the used research methods. Xie et al. [44] summarized the existing solution methods for the FJSP and classified them into exact algorithms, heuristics, and meta-heuristics. They also introduced real-world applications of the FJSP and analyzed the development trends of the manufacturing industry. A comprehensive literature review of the integrated FJSP and its extensions were presented in the work of [25] where they analyzed 140 research paper.

The paper by [30] presents a hybrid artificial bee colony (hyABC) algorithm to minimise the total flowtime for a FJSP with overlapping in operations. They also developed a modified migrating birds optimization algorithm (MBO) is developed and integrated into the search process. The comparisons with other recent algorithms identify the effectiveness of the proposed hyABC.

The research of [47] presents a two-level particle swarm optimization algorithm for the flexible job shop scheduling problem. The upper level handles the operations-to-machines mapping, while the lower level handles the ordering of operations on machines.

In [1], author develops a hybrid meta-heuristic algorithm, including a particle swarm optimization (PSO) procedure and elements of tabu search (TS) meta-heuristic to solve flexible job shop scheduling problems (FJSSPs). Twelve benchmark test examples from different reference sources are experimentally tested to demonstrate the performance of the algorithm.

The paper by [42] focuses on the dynamic flexible job shop scheduling problem considering the implicit and explicit deterioration effect (DFJSP-DE). They formulated a multi-objective optimization model: makespan, energy consumption, and stability of rescheduling solutions. The results of three numerical experiments show that the proposed approach can solve DFJSP-DE effectively.

The authors in [30] address the distributed flexible job shop scheduling problem (DFJSP) with minimizing energy consumption using a novel mixed integer linear programming (MILP) model to solve small-scaled problems to optimality. For large-size problems, they propose an efficient hybrid shuffled frog-leaping algorithm (HSFLA). The performance of these algorithms was confirmed by numerical experiments.

The authors in [39] propose a multi-population GA with an Erdos and Renyi model (ER) network (MPGA-ER) and test the performance according to the total individual number (TIN). We study how the subpopulation number and size affect the propagation rate of advantageous genes. They use a parameter-optimized MPGA-ER to solve for more FJSP instances and demonstrate its effectiveness.

The paper by [24] establishes a mathematical optimization model of flexible job shop scheduling with adjustment time based on minimizing the makespan. The improved genetic algorithm is used to solve the proposed model and achieve rapid optimization of the solution process.

In [32], researchers present a Q-learning rescheduling approach to the flexible job shop problem merging energy and productivity objectives in the context of machine collapse. They adopted a genetic algorithm to generate the initial predictive schedule. Then, they used rescheduling strategies to handle machine failures where the optimal schedule was selected by a multi-objective Q-learning algorithm that minimizes the makespan and the energy consumption variation at the same time.

The paper by [31] proposes a scheduling rule-based surrogate assisted simulation–optimization approach for solving a combinatorial optimization problem related to a realistic FJSSP. The approach is applied to a highly automated flexible robotized manufacturing system (FMS) integrating different realistic and representative constraints to the classical FJSSP.

In [21], authors study flexible job shop scheduling problems with arbitrary precedence graphs. They proposed mixed integer and constraint programming models and evolutionary algorithms to solve large-scale problems. Overall, 59 new best solutions and 61 new lower bounds are produced for a total of 228 benchmark problem instances of the literature. This work was expanded with nonlinear routes or equivalently with arbitrary precedence graphs [20].

The authors in [27] suggest a hierarchical and distributed architecture to resolve the problem of dynamic flexible job shop scheduling. They used the double deep Q-network algorithm in training scheduling agents to trace the link between production information and the objectives of scheduling with real-time scheduling decisions according to job arrivals. They also improved the learning and scheduling efficiency by developing a surrogate reward-shaping technique. The same technique was used by [6] in addition to a soft ε-greedy behavior policy which was designed according to the scale of the problem.

A recent work by [26] adopted an improved immune genetic algorithm (IGA) based on greedy thought combined with local scheduling rules to solve the flexible job shop scheduling problem and a batch process (BP) problem. In the flexible job shop part, the greedy optimal solution is obtained through the greedy thought. The results show that the proposed method can effectively solve such problems.

Recently, [28] proposed a dynamic self-learning artificial bee colony (DSLABC) optimization algorithm to solve the dynamic flexible job shop scheduling problem (DFJSP). They combine the Q-learning algorithm and the traditional artificial bee colony (ABC) to form the self-learning artificial bee colony (SLABC) algorithm. Then, the specific method of dynamic scheduling is determined, and the DSLABC algorithm is proposed. Solving the Brandimarte instances, it is proved that the convergence accuracy of the SLABC algorithm is higher.

Many studies have attempted to resolve job shop scheduling problems or flexible job shop scheduling problems through the enhancement of genetic algorithms. Somehow, in most of these studies, the genetic algorithms were improved mostly through making alteration of the pheromone update mechanism. While it is true that such alteration can produce better search speed and solution efficiency, premature convergence may result if the pheromone feedback of the best path is strengthened too much. Hence, for resolving FJSP, this study proposes the application of the hybridized GA and SA methods for improving the pure genetic algorithm (GA). In fact, the outcomes appear to be nearer or equivalent to the global optimum. In demonstrating the effectiveness of the proposed method, this study carries out numerical experiments with the use of benchmark problems. The obtained outcomes provide validation to the quality of the proposed approach.

Even though there are many similar works, this current study differs in terms of the application order of GA and SA, chromosome modeling, improved crossover and mutation operators, and adaptive simulated annealing. Also, we contribute to the existing studies by the application of our algorithm on a large number of instances (162).

In essence, what is left of this chapter is structured as follows: Sect. 2 illustrates the representation of the solution for FJSP; Sect. 3 discusses the meta-heuristic algorithm proposed; Sect. 4 provides the experimental test, comparison, and discussion; and Sect. 5 concludes the chapter with suggestions for future research.

2 The Solution Representation

The vector or the individual representation is introduced in this study using an example of three machines' problem containing three tasks. In this regard, the 3x3 FJSP displayed in Table 2 is referred. Here, job 1 (J_1) contains three operations (O_{11}, O_{12}, *and* O_{13}), job 2 (J_2) contains three operations (O_{21}, O_{22}, *and* O_{23}), and job 3 (J_3) contains two operations (O_{31} *and* O_{32}). Based on that, one potential schedule could be as follows: $O_{21}, O_{11}, O_{22}, O_{12}, O_{13}, O_{31}, O_{23}, and\ O_{32}$.

Table 2 Example (3*3)

		M_1	M_2	M_3
J_1	O_{11}	1	×	4
	O_{12}	×	×	2
	O_{13}	3	×	×
J_2	O_{21}	×	1	1
	O_{22}	2	×	×
	O_{23}	×	1	×
J_3	O_{31}	×	6	3
	O_{32}	4	×	×

Number of operations	1	2	3	4	5	6	7	8
Assignment of operations V_1	1	2	1	2	1	2	3	1

Sequence of operations	O_{11}	O_{21}	O_{12}	O_{22}	O_{31}	O_{13}	O_{23}	O_{32}
Oder of operations V_2	1	4	2	5	7	3	6	8

Fig. 1 Assignment and sequence vector

This chapter employs the system of encoding proposed by Gao. In this system, the representation generally includes two parts, namely machine assignment vector V and operation sequence vector. Both parts are explained as follows:

1. Machine assignment vector V_1, $V_1(s)$ signifies the chosen machine for the operation specified at point s. In Fig. 1, for instance, position 4 denotes O_{21}, while V_1 (4) denotes the machine allocated for O_{21}.
2. Operation sequence vector V_2 signifies the practicable sequence that matches V_1. Figure 1, for instance, can be transformed into a list of order sequence. In this representation, each conceivable chromosome constantly signifies a practicable sequence of operation, and the space of coding appears to be smaller than the space of coding of permutation representation [13].

3 The Proposed Hybrid Algorithm

In minimizing makespan in this kind of problem of scheduling, this study demonstrates the application of an effective hybrid meta-heuristic grounded upon genetic algorithm (GA) and a simulated annealing (SA) procedure.

3.1 *Simulated Annealing*

Simulated annealing (SA) by [22] encompasses a meta-heuristic grounded upon the similarity between the solid process of annealing and the resolution of combinatorial optimization problems [23, 40]. There are several decreasing temperatures in SA, and there are a number of iterations contained in each temperature. In the use of SA, there are a few steps. The first step is to choose the initial temperature T_0. The beginning solution X_0 is also selected randomly. Then, computation is made to determine the makespean value. For this purpose, the present solution X_{curent} (i.e., X_0 in this case) is employed. Here, the aim is to minimize the makespean C_{max}. This is followed by the production of a new solution X_{new} from the adjacent of $X_{current}$. Computation is then made to the makespean of X_{new} and $X_{current}$. In this regard, X_{new} will be accepted if C_{max} (X_{new}) is lower than C_{max} ($X_{current}$), or else, it will be accepted only if it fulfills the criterion of Metropolis. The criterion of Metropolis is grounded upon the probability of Boltzman. The criterion of Metropolis posits that if the dissimilarity between $C_{\max}$ (X_{new}) and $C_{\max}$ (X_{current}) (i.e., ΔE) is equivalent to or bigger than zero, a random number p in [0, 1] is produced from a distribution. If Eq. (1) is fulfilled, the newly produced solution is recognized as the present solution.

$$p \leq e^{\left(-\frac{\Delta E}{T}\right)} \tag{1}$$

At each temperature, the amount of produced new solutions tallies with the amount of iteration at the temperature is limited by the condition of termination T_{f}. The condition of the termination could be as unpretentious as the specified number of iterations. Then, following the completion of all iterations at a temperature, the temperature would be reduced. Here, the temperature updating rule will be referred. Prior to shifting to the subsequent temperature, all mandatory iterations will have to be finished at the revised (and reduced) temperature. This process would be run over and over again until it fulfills the stopping criterion, which could be T_{f}. The SA outcome is linked to the number of iterations at each temperature and the rapidity of dropping temperature. Accordingly, this chapter proposes the temperature updating rule as expressed below:

$$T = \beta T_0 \tag{2}$$

From the above expression, T_0 signifies the beginning temperature, that is, the ratio of cooling, which governs the cooling speed. The approach used in this study employs swapping as well as the insertion neighborhood function. An adjacent solution is produced in each iteration. For this purpose, one of these neighborhoods is randomly selected.

3.2 *Genetic Algorithm*

Genetic algorithms (GAs) encompass the mixture of selection, recombination, and mutation for evolving a solution to a problem, and these algorithms can be employed in the resolution of diverse types of problems related to scheduling [8, 9]. Studies that demonstrate the adaptation of genetic algorithms to the flexible job shop scheduling problem offers valuable relevant information [19]. In addition, expansive literature on the successful utilization of GAs to this problem is available for scrutiny. Relevantly, the application of the proposed algorithm encompasses several steps as detailed below.

Initialization In a genetic algorithm, random production of the initial population is a common task. Somehow, it has become a trend nowadays to include several good individuals in the population via several effective heuristics or by certain dispatching rules. The use of this approach guarantees a swifter convergence to good solutions.

For the purpose of this study, two initial populations are proposed as follows: pop_r and pop_l. In pop_r, the operations are randomly allotted to different machines, while in pop_l, good individuals are obtained from local search procedure as illustrated below:

Algorithm: Local Search Procedure
Begin
Select the best individual X^r_{best} in pop_r;
Loop
Randomly generate N individual in the neighborhood of X^r_{best};
Calculate the makspean values of the generated new individuals;
Select the individual $X_{current}$ with the optimal value among the makspean ones of the new individuals;
If $\boldsymbol{C_{max}}\left(X_{current}\right) < \boldsymbol{C_{max}}\left(X^r_{best}\right)$ ***Then***;
Add $X_{current}$ to pop_1;
$X^r_{best} \leftarrow X_{current}$;
End
End

Fitness Evaluation The makespan is calculated for each chromosome within the present population.

Selection The best chromosomes are identified at each iteration for the regeneration of linear ranking. Then, based on the distribution of probability, the parents are selected.

$$p_i = \frac{2r_i}{N_p\left(N_p + 1\right)} \tag{3}$$

From the expression above, p_i denotes the probability of selecting the ith individual within the order of the rank. Arrangement is made for the individuals based on their fitness, and a rank $r_i =$ is allotted to each, whereby N_p denotes the size of the population. The best individual is assigned the N_p rank, and the worst individual is assigned rank 1.

Crossover For problems of scheduling, the literature presents a number of crossover operators. The crossover operator generally produces two children, from the two chosen progenitors. For the purpose of this study, the one-point crossover operator has been chosen, and it is employed with a P_c probability to two distinctive individuals chosen via linear ranking, and this produces two children (Fig. 2).

Application of the Simulated Annealing Algorithm The SA algorithm is provisionally run following the permission of a new solution. Accordingly, the child makespan containing the best result $C^{\text{best}}_{\text{max}}$ is evaluated, multiplied by $(1 + \alpha)$. Here, α denotes a parameter algorithm established via experimentation. The SA algorithm would be applied if the child evaluation is lower than the $(1 + \alpha)$ $C^{\text{best}}_{\text{max}}$. The procedure will be run again and again and will stop until probable improvement

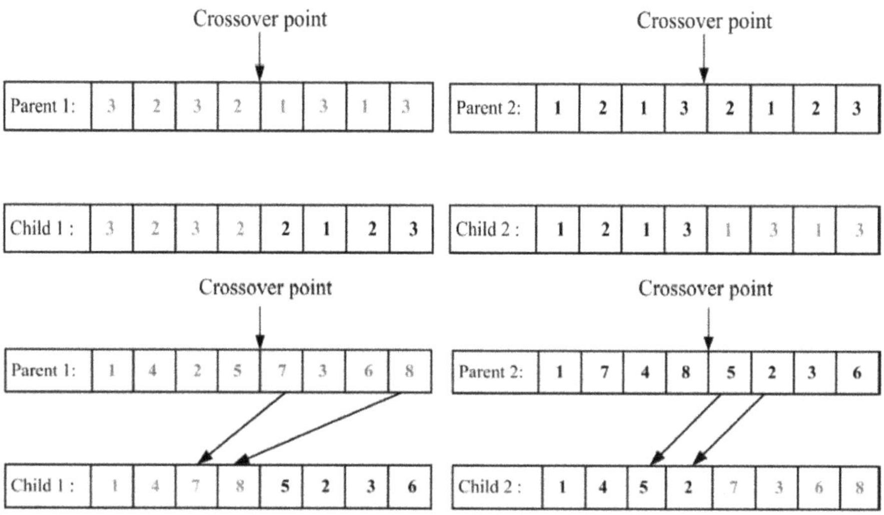

Fig. 2 Crossover of the assignment vector V_1 and the sequencing vector V_2

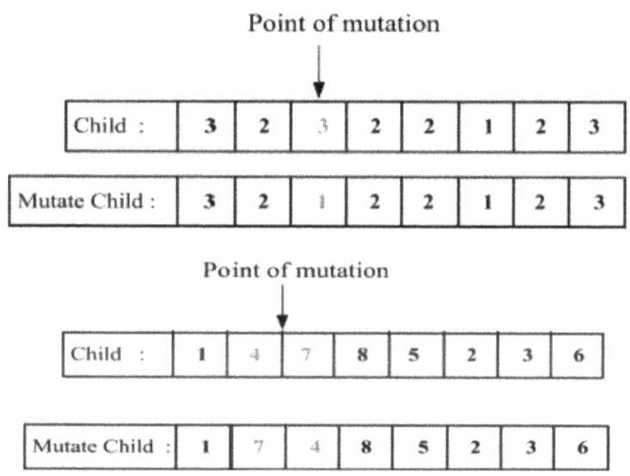

Fig. 3 Mutation of the assignment vector V1 and the sequencing vector V2

no longer occurs. Hence, the SA algorithm is employed as a way to escape from local optima.

Replacement The proposed algorithm provides that the descendant will be accepted as part of the population provided that it is superior as opposed to the worst population members or is unique from those already belonging to the population. The attained population gradually progresses to better makespan values. At the same time, it carries different solutions, and this assists in diversity preservation.

Mutation Mutation allows genetic algorithms to explore a bigger area of the solution space. In this regard, random genes or chromosome changes are introduced. When applying the suggested algorithm, a P_m probability is replaced by the assignment of an operation in V_1. Otherwise, two neighboring operations are transposed from two dissimilar jobs in V_2 (Fig. 3).

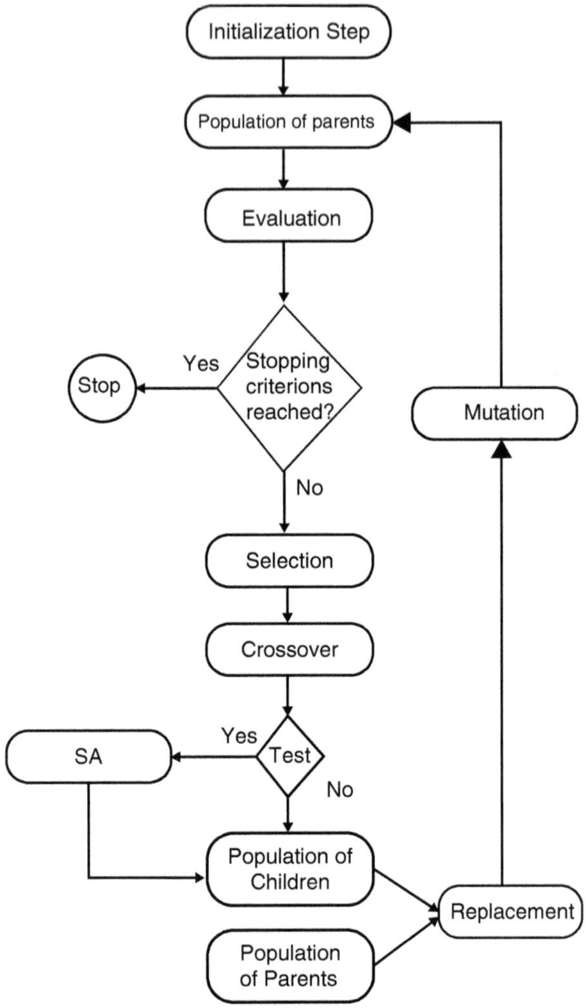

Fig. 4 The main steps of the proposed GA-SA

The proposed GA-SA procedure is summarized in Fig. 4.

4 Experimental Results

The effectiveness as well as performance of the proposed GA-SA algorithm is proven in this study. For the purpose, the procedure of the algorithm was implemented in C++ using a personal computer running on PC Intel (R) Core(TM) i5-6200U CPU @ 2.30GHz and 4.00 GB of RAM memory. The input entails four representative classes of instances grounded upon practical data, and each instance

Table 3 Problem instances

Instances	Num	Flexibility			n	M	n_i	N
		Total	Partial	Range				
Kacem data (Ka.data)	3	2	1	–	8–15	8–10	2–4	15–56
Fattahi data (Fa.data)	20	0	20	–	2–12	2–8	2–4	4–48
Brandimarte data (Br.data)	10	0	10	1.43–4.10	10–20	4–15	5–15	55–240
Hurink data (Hu.data)	129	0	129	1.15–7.50	6–30	5–15	5–15	36–300

has the following attributes: number of jobs (n), number of machines (m), number of operations (n_i) in job J_i, and the overall number of operations (N). Due to the number of flexibility and nature of the instances, many runs are carried out on the same problem instance so that meaningful outcomes can be achieved.

After ten runs of GA-SA from different population size have been carried out, the best solutions for each problem are identified and selected. Then, a comparison is made between the proposed algorithm and the advanced algorithms, performance-wise. Accordingly, Table 3 presents the features of problem instances in summary form. In the first column, the dataset is presented, while the second column shows the number of instances allocated for each class, whereas the third column shows the number of instances with overall flexibility. The number of instances with partial flexibility is presented in the fourth column, while the flexibility range for each operation is displayed in the fifth column. The minimal and maximal number of jobs n, number of machines m, number of operations n_i within job J_i, and total number of operations N are placed in their respective column.

Considering the major impact of the chosen parameter values on the achieved results, comprehensive initial computational tests are performed in this study. For this purpose, the parameter tuning is employed, and the details are as follows: For GA, $p_c = 1$, $p_m = 0.05$, and the number of maximal generation is 1000. For SA, $T_0 = -1/\log(0.01)$, $T_f = -1/\log(0.001)$, $p = 0.01$, and $\beta = 0.999$. Determined by the difficulty of the problem, the range of the GA population size is from 500 to 4000.

Kacem data is the first investigated dataset. Accordingly, Table 4 shows the comparison between the outcomes obtained from the proposed algorithm and those obtained using the localization (AL + GA) as was presented in [19], the hybrid algorithm of particle swarm (PSO + SA) as was presented in [43], the multistage genetic algorithm (moGA) as was presented in [50], the hybrid genetic algorithm (hGA) as was presented in [13], the parallel variable neighborhood search algorithm (PVNS) as was presented in [46], and also the artificial immune algorithm (AIA) as was presented in [2].

In Table 4, the first column shows the instances under studding, while the second through the sixth column displays the outcomes of the aforementioned approaches.

Table 4 Results on Kacem data

Instances	Size n * m	AL + GA	PSO + SA	Mo GA	HGA	PVNS	AIA			GA-SA		
							C_{max}	Pop. size	CPU	C_{max}	Pop. size	CPU
Ka08	8 * 8	15	15	15	15	14	14	400	0.76	14	500	0.8
Ka10	10 * 10	7	7	7	7	7	7	2000	8.97	7	500	0.97
Ka15	15 * 10	24	12	11	11	12	11	5000	109.2	11	500	6.04

On the other hand, the seventh through the ninth column, respectively, highlights the best makespan, the population and the computational time (second) of the artificial immune algorithm (AIA) as was presented in [2]. Lastly, the remaining three columns show the outcomes obtained from the proposed algorithm. As can be construed, the proposed GA-SA algorithm produces similar best $C_{\max}$ values for all instances. In addition, the algorithm has more speed as opposed to the artificial immune algorithm (AIA) introduced in [2].

The second dataset comprises Fattahi data problems. In Table 5, the proposed GA-SA algorithm is compared to the algorithms proposed by [2, 12], and the multi-population GA with an ER network (MPGA-ER) in [39]. Here, 20 FJSP problem instances obtained from Fattahi [18] are used. The beginning column contains the name of the instance, followed by the succeeding column which contains the amount of both jobs and machines for the instances. The ensuing column, which is the third column, shows the best lower bound [12]. In the fourth, fifth, and sixth columns, the best makespan over ten runs of the proposed GA-SA algorithm, the CPU time as well as the relative deviation (dev (%)) is respectively displayed. Further, the seventh, eighth, and ninth columns, respectively, display the makespan, the CPU time, and the relative deviation (dev (%)) generated by the artificial immune algorithm (AIA) from [2]. The rest of the columns show the best results in addition to the relative deviation of two algorithms introduced in [12]. The expression of relative deviation is as shown below:

$$\text{dev} = \left(\left(C_{\max}^{\text{other}} - C_{\max}^{\text{best}}\right) / C_{\max}^{\text{best}}\right) \times 100 \tag{4}$$

where $C_{\max}^{\text{best}}$ and $C_{\max}^{\text{other}}$, respectively, signify the best makespan achieved using the proposed and the reference algorithms. In Table 5, the ending row signifies the resultant average deviation of all algorithms. Meanwhile, Table 6 shows the best outcomes obtained using the proposed GA-SA algorithm on Fattahi data compared to four hierarchical approaches (i.e., (hybrid simulated annealing (HSA)/simulated annealing (SA), hybrid simulated annealing (HSA)/Tabu search (TS), Hybrid Tabu search (HTS)/Tabu search (TS), and Hybrid Tabu search (HTS)/simulated annealing (SA)) proposed by [12].

As can be deduced from the outcomes displayed in Tables 5 and 6, the results from GA-SA are the most superior, and all the average deviations appear to be positive. Furthermore, as can be observed from the results shown in Table 7, the proposed GA-SA algorithm supersedes the approach proposed by Bagheri in 5 out of 20 problems based on makespan (Cmax). In addition, the proposed GA-SA algorithm supersedes five algorithms of Fattahi in nine instances at least.

Table 8 presents the mean relative error (MRE) of the best outcomes generated using GA-SA and seven algorithms presented by [12]. The MRE shown in Table 8 represents the average mean relative error for all instances. Accordingly, the starting column displays the dataset, while the second column presents the average number of alternative machines for each operation. The expression of relative error (RE) is shown below:

Table 5 Comparing GASA with other algorithms using Fattahi instances

Instances	Size n * m	*LB*	GASA		AIA			ISA		ITS		MPGA-ER	
			$C_{\max}$	CPU	$C_{\max}$	CPU	Dev (%)	$C_{\max}$	Dev (%)	$C_{\max}$	Dev (%)	$C_{\max}$	Dev (%)
SFJS1	2 *2	66	66	0, 00	66	0, 03	0,00	66	0,00	66	0,00	66	0,00
SFJS2	2 * 2	107	107	0, 00	107	0, 03	0,00	107	0,00	107	0,00	107	0,00
SFJS3	3 * 2	221	221	0, 00	221	0, 04	0,00	221	0,00	221	0,00	221	0,00
SFJS4	3 * 2	355	355	0, 00	355	0, 04	0,00	355	0,00	390	9,86	355	0,00
SFJS5	3 * 2	119	119	0, 00	119	0, 04	0,00	119	0,00	137	15,13	119	0,00
SFJS6	3 * 3	320	320	0, 00	320	0, 04	0,00	320	0,00	320	0,00	320	0,00
SFJS7	3 * 5	397	397	0, 00	397	0, 04	0,00	397	0,00	397	0,00	397	0,00
SFJS8	3 * 4	253	253	0, 00	253	0, 05	0,00	253	0,00	253	0,00	253	0,00
SFJS9	3 * 3	210	210	0, 00	210	0, 05	0,00	215	2,38	215	2,38	210	0,00
SFJS10	4 * 5	516	516	0, 00	516	0, 05	0,00	516	0,00	617	19,57	516	0,00
MFJS1	5 * 6	396	468	0.03	468	9, 23	0,00	488	4,27	548	17,09	468	0,00
MFJS2	5 * 7	396	448	0, 09	448	9, 35	0,00	478	6,70	457	2,01	446	−0.42
MFJS3	6 * 7	396	466	8, 48	468	10, 06	0,43	599	28,54	606	30,04	466	0,00
MFJS4	7 * 7	496	564	0, 06	554	10, 54	−1,77	703	24,65	870	54,26	554	−1,77
MFJS5	7 * 7	414	514	0, 62	527	10, 61	2,53	674	31,13	729	41,83	514	0,00
MFJS6	8 * 7	469	634	0, 75	635	22, 18	0.16	856	35,02	816	28,71	634	0,00
MFJS7	8 * 7	619	879	10, 63	879	24, 82	0,00	1066	21,27	1048	19,23	879	0,00
MFJS8	9 * 8	619	884	3, 07	884	26, 94	0,00	1328	50,23	1220	38,01	884	0,00
MFJS9	11 * 8	764	1077	17, 50	1088	30, 76	1,02	1148	6,59	1124	4,36	/	/
MFJS10	12 * 8	944	1208	20, 29	1267	30, 94	4,88	1546	27,98	1737	43,79	/	/
Average deviation							***0,36***		***11,94***		***16,31***	***−0,12***	
Relative deviation							***3.58***		***1.31***		***1.08***	***3.48***	

Table 6 Comparison of GA-SA with Fattahi algorithms

Instances	Size n * m	*LB*	GASA	HAS/SA		HAS/TS		HTS/TS		HTS/SA	
			$C_{\max}$	$C_{\max}$	Dev (%)	$C_{\max}$	Dev (%)	$C_{\max}$	Dev (%)	$C_{\max}$	Dev (%)
SFJS1	2 * 2	66	66	66	0,00	66	0,00	66	0,00	66	0,00
SFJS2	2 * 2	107	107	107	0,00	107	0,00	107	0,00	107	0,00
SFJS3	3 * 2	221	221	221	0,00	221	0,00	221	0,00	221	0,00
SFJS4	3 * 2	355	355	355	0,00	355	0,00	355	0,00	355	0,00
SFJS5	3 * 2	119	119	119	0,00	119	0,00	119	0,00	119	0,00
SFJS6	3 * 3	320	320	320	0,00	320	0,00	320	0,00	320	0,00
SFJS7	3 * 5	397	397	397	0,00	397	0,00	397	0,00	397	0,00
SFJS8	3 * 4	253	253	253	0,00	253	0,00	253	0,00	256	1,19
SFJS9	3 * 3	210	210	210	0,00	210	0,00	210	0,00	210	0,00
SFJS10	4 * 5	516	516	516	0,00	516	0,00	516	0,00	516	0,00
MFJS1	5 * 6	396	468	479	2,35	491	4,91	469	0,21	469	0,21
MFJS2	5 * 7	396	448	495	10,49	482	7,59	482	7,59	468	4,46
MFJS3	6 * 7	396	466	553	18,67	538	15,45	533	14,38	538	15,45
MFJS4	7 * 7	496	564	656	16,31	650	15,25	634	12,41	618	9,57
MFJS5	7 * 7	414	5[illegible]4	650	26,46	662	28,79	625	12,60	625	21,60
MFJS6	8 * 7	469	634	762	20,19	785	23,82	717	13,09	730	15,14
MFJS7	8 * 7	619	879	1020	16,04	1081	22,98	964	9,67	947	7,74
MFJS8	9 * 8	619	884	1030	16,52	1122	26,92	970	9,73	922	4,30
MFJS9	11 * 8	764	1077	1180	9,56	1243	15,41	1105	2,60	1105	2,60
MFJS10	12 * 8	944	1208	1538	27,32	1615	33,6	1404	16,23	1384	14,57
Average deviation					***8,20***		***9,74***		***5,38***		***4,84***
Relative deviation					***1.21***		***1.22***		***1.25***		***1.4***

Table 7 GA-SA versus seven approaches from Bagheri and Fattahi on Fattahi data

Approach	Num of instances	Better C_{max}	Equal C_{max}	Worse C_{max}
AIA	*20*	*5*	*14*	*1*
ISA	*20*	*11*	*9*	*0*
ITS	*20*	*14*	*6*	*0*
HSA/SA	*20*	*10*	*10*	*0*
HSA/TS	*20*	*10*	*10*	*0*
HTS/TS	*20*	*10*	*10*	*0*
HTS/SA	*20*	*11*	*9*	*0*
MPGA-ER	*18*	*0*	*16*	*2*

Table 8 Mean relative error (%) on Fattahi data

Alt.	GA-SA	AIA	HSA/SA	HSA/TS	HTS/TS	HTS/SA	ISA	ITS	MPGA-ER
2,15	13,78	14,27	24,29	26,48	20,61	19,9	29,27	33,84	11.35

Table 9 Comparison presented GA-SA with AIA and PSO on Brandimarte data

Instances	Size	LB	GA-SA		AIA		PSO
	n * m		C_{max}	CPU	C_{max}	CPU	C_{max}
MK01	10 ×6	36	40	3,73	40	97,21	41
MK02	10 × 6	24	26	22,26	26	103,46	26
MK03	15 × 8	204	204	1,00	204	247,37	207
MK04	15 × 8	48	60	8,10	60	152,07	65
MK05	15 × 4	168	173	9,10	173	171,95	171
MK06	10 × 15	33	61	219,67	63	245,62	61
MK07	20 × 5	133	**139**	16,65	140	161,92	173
MK08	20 × 10	523	523	7,79	523	392,25	523
MK09	20 × 10	299	307	74,31	312	389,71	307
MK10	20 × 15	165	**210**	298,31	214	384,54	312

$$\mathrm{RE} = \left[\left(C_{max}^{best} - \mathrm{LB}\right)/\mathrm{LB}\right] \times 100 \tag{5}$$

whereC_{max}^{best} signifies the best makespan of the described algorithm, while LB denotes the best recognized lower bound.

Table 9 shows the class of instances of Brandimarte. As shown, the starting column displays the names of the instance, followed by the second column that contains the size of the instance. The third column displays the lower bound. The makespan value for each algorithm as well as the CPU time is presented for each algorithm. As highlighted, the outcomes of the proposed algorithm appear to be similar or superior to those obtained from AIA and the PSO algorithm proposed by [33].

Table 10 presents the computation of the relative deviation for comparison between the proposed algorithm and five other algorithms (i.e., AIA, GA Chen, GA Gao, GA Pezzella, and PVNS). The first column in Table 10 contains the names of

Table 10 Results on Brandimarte data

Instances	Size	LB	GA-SA	AIA		GA Chen		GA Gao		GA Pezzella		PVNS	
	n * m	C_{max}		C_{max}	Dev (%)	C_{max}	Dev (%)	C_{max}	Dev (%)	C_{max}	Dev (%)	C_{max}	Dev (%)
MK01	10 × 6	36	40	40	0,00	40	0, 00	40	0,00	40	0,00	40	0, 00
MK02	10 × 6	24	26	26	0,00	29	11, 54	28	0,77	26	0,00	26	0, 00
MK03	15 × 8	204	204	204	0,00	204	0, 00	204	0,00	204	0,00	204	0, 00
MK04	15 × 8	48	60	60	0,00	63	5, 00	61	0,17	60	0,00	60	0, 00
MK05	15 × 4	168	173	173	0,00	181	4, 62	176	0,17	173	0,00	173	0, 00
MK06	10 × 15	33	61	63	3,28	60	−1, 64	62	0,16	63	3,28	60	−1, 64
MK07	20 × 5	133	139	140	0,72	148	6, 47	145	0,43	139	0,00	141	1, 44
MK08	20 × 10	523	523	523	0,00	523	0, 00	523	0,00	523	0,00	523	0, 00
MK09	20 × 10	299	307	312	1,63	308	0, 33	310	0,10	311	1,30	307	0, 00
MK10	20 × 15	165	210	214	1,90	212	0, 95	216	0,29	212	0,95	208	−0, 95
Average deviation					0,75		2, 73		0,21		0,55		−0, 12
Relative deviation					1.53		1.5		1.15		1.93		6.82

Table 11 Mean relative error (%) on Brandimarte data

Instances	GA-SA	AIA	PVNS	GA Chen	GA Jia	GA Pezzella
Brandimarte data	*16,67*	*17,76*	*16,39*	*19,55*	*19,11*	*17,53*

Table 12 GA-SA versus five approaches (AIA, GA Chen, GA Gao, GA Pezzella, and PVNS) on Brandimarte data

Approach	Num C_{max}	Better C_{max}	Equal C_{max}	Worse C_{max}
AIA	10	4	6	0
GA Chen	10	6	3	1
GA Gao	10	7	3	0
GA Pezzella	10	3	7	0
PVNS	10	1	7	2

Table 13 Mean relative error on Hurink data

Instances	Flexibility	GA-SA	AIA	PVNS	GA Chen	GA Jia	GA Pezzella
Edata	1,15	4,22	6,83	3,86	5,59	9,01	6,00
Rdata	2	2,28	3,98	1,88	4,41	8,34	4,42
Vdata	4,31	0,31	1,29	0,42	2,59	3,24	2,04

the instance, the second column displays the size, the third column shows the lower bound, and the fourth column shows the acquired value of makespan.

The value of makespan for each algorithm is documented, while the relative deviation is computed. Meanwhile, the ending row shows the established average relative deviation. As can be seen from the table, the proposed algorithm achieved positive average relative deviation, implying the superiority of the algorithm over others. A negative deviation was achieved for the last algorithm.

Table 11 presents the comparison of the mean relative error (MRE) of the proposed algorithm and that of other algorithms (i.e., AIA, PVNS, GA Chen, GA Jia, and GA Pezzella) based on Brandimarte data. As should be noted, the algorithm yielding the smallest value is considered the best algorithm.

As can be viewed in Table 12, the starting column contains the different approaches, and in the next column, the number of instances are displayed. Based on the computation of the number of instances, the makespan of GA-SA appears to be better, equivalent or worse than the makespans reported in [2, 13, 35, 46].

Finally, we apply our algorithm to the data of Hurink [17]. We use 129 instances split into three parts: Edata, Rdata, and Vdata classified according to the degree of flexibility. Regarding the large number of instances, we reported only the mean relative error.

The average of relative errors computation for the Hunrik data is presented in Table 13. As opposed to other algorithms for Edata and Rdata, the proposed method is in second place. However, the proposed method ranked first in the comparison of the Vdata results.

Table 14 t-test: paired two samples for means (Fattahi instances) [(*) significant at 5%]

	HAS/SA	HAS/TS	HTS/SA	HTS/TS	AIA	ISA	ITS	MPGA-ER
t-stat	−3.187	−3.083	−3.174	−3.043	−1.236	−3.054	−3.353	1.190
P value	0.002*	0.003*	0.003*	0.003*	0.117	0.003*	0.002*	0.125

Table 15 t-test: paired two samples for means (Brandimarte instances) [(*) significant at 5%]

	AIA	PSO	GA Chen	GA Gao	GA Pezzella	PVNS
t-stat	−2.025	−1.388	−2.299	−3.025	−1.809	0.318
P value	0.037*	0.09	0.023*	0.007*	0.05*	0.379

To check the robustness of our results, we apply a paired two samples for means. The test is performed for all the algorithms using the instances of Fattahi and Brandimarte. The hypotheses for one tail test are:

H0: $Cmax_{algorithm} \leq Cmax_{GASA}$
H1: $Cmax_{algorithm} > Cmax_{GASA}$

A significant test at the level of 5% indicates that our algorithm Genetic algorithm simulated annealing (GASA) outperforms the other algorithms in minimizing makespan. The results are summarized in Tables 14 and 15.

The tests above confirm the results issued from the comparison between GASA and other algorithms, where our algorithm outperforms all the algorithms using Fattahi data except ISA and MPGA-ER. Under Brandimarte instances, only PSO and PVNS give better results in minimizing the makespan.

5 Conclusion

This study demonstrates the application of a hybrid genetic algorithm with simulated annealing (GA-SA) for the resolution of the flexible job shop scheduling problem (FJSSP), with the specific purpose of makespan minimization. Accordingly, a huge set of instances with diverse sizes (small, medium, and large) was tested for performance comparison of the proposed algorithm. Notably, good results were obtained. Importantly, future work should explore multi-objective FJSSP associated with the fuzzy due date and energy consumption. Also, dynamic scheduling for FJSP may also be an area of interest. This is because in the real-world systems of manufacturing, unplanned events often occur, for instance, new arrivals of jobs and machine failure.

References

1. T. Asen, Particle swarm optimization and tabu search hybrid algorithm for flexible job shop scheduling problem – Analysis of test results. Cybernitic Inf. Technol. **19**(4), 26–44 (2019)
2. A. Bagheri, M. Zandieh, I. Mahdavi, M. Yazdani, An artificial immune algorithm for the flexible job-shop scheduling problem. Futur. Gener. Comput. Syst. **26**, 533–541 (2010)
3. J.W. Barnes, J.B. Chambers, Solving the job shop scheduling problem with tabu search. IIE Trans. **27**, 257–263 (1995)
4. P. Brandimarte, Routing and scheduling in a flexible job shop by tabu search. Ann. Oper. Res. **41**, 157–183 (1993)
5. P. Brucker, R. Schlie, Job-shop scheduling with multi-purpose machines. Computing **45**, 369–375 (1990)
6. J. Chang, D. Yu, Y. Hu, W. He, H. Yu, Deep reinforcement learning for dynamic flexible job shop scheduling with random job arrival. Processes **10**(4), 760 (2022)
7. A.I. Chaudhry, A.A. Khan, A research survey: Review of flexible job shop scheduling techniques. Int. Trans. Oper. Res. **23**, 551–591 (2016)
8. R. Cheng, M. Gen, Y. Tsujimura, A tutorial survey of job-shop scheduling problems using genetic algorithms – I. Represent. Comput. Ind. Eng. **30**, 983–997 (1996)
9. R. Cheng, M. Gen, Y. Tsujimura, A tutorial survey of job-shop scheduling problems using genetic algorithms, part II: Hybrid genetic search strategies. Comput. Ind. Eng. **36**, 343–364 (1999)
10. S. Dauzère-Pérès, J. Paulli, An integrated approach for modeling and solving the general multiprocessor job-shop scheduling problem using tabu search. Ann. Oper. Res. **70**, 281–306 (1997)
11. L. De Giovanni, F. Pezzella, An improved genetic algorithm for the distributed and flexible job-shop scheduling problem. Eur. J. Oper. Res. **200**, 395–408 (2010)
12. P. Fattahi, M.S. Mehrabad, F. Jolai, Mathematical modeling and heuristic approaches to flexible job shop scheduling problems. J. Intell. Manuf. **18**, 331–342 (2007)
13. J. Gao, L. Sun, M. Gen, A hybrid genetic and variable neighborhood descent algorithm for flexible job shop scheduling problems. Comput. Oper. Res. **35**, 2892–2907 (2008)
14. M.R. Garey, D.S. Johnson, R. Sethi, The complexity of flowshop and jobshop scheduling. Math. Oper. Res. **1**, 117–129 (1976)
15. N.B. Ho, J.C. Tay, GENACE: An efficient cultural algorithm for solving the flexible job-shop problem, in *Evolutionary Computation, CEC 2004. Congress on 2004* (2004), pp. 1759–1766
16. N.B. Ho, J.C. Tay, E.M.-K. Lai, An effective architecture for learning and evolving flexible job-shop schedules. Eur. J. Oper. Res. **179**, 316–333 (2007)
17. J. Hurink, B. Jurisch, M. Thole, Tabu search for the job-shop scheduling problem with multi-purpose machines. Oper.-Res.-Spektrum **15**, 205–215 (1994)
18. I. Kacem, S. Hammadi, P. Borne, Approach by localization and multiobjective evolutionary optimization for flexible job-shop scheduling problems. IEEE Trans. Syst. Man Cybern. Part C Appl. Rev. **32**, 1–13 (2002)
19. I. Kacem, S. Hammadi, P. Borne, Pareto-optimality approach for flexible job-shop scheduling problems: hybridization of evolutionary algorithms and fuzzy logic. Math. Comput. Simul. **60**, 245–276 (2002)
20. G.A. Kasapidis, S. Dauzère-Pérès, D.C. Paraskevopoulos, P.P. Repoussis, C.D. Taantilis, On the multiresource flexible job-shop scheduling problem with arbitrary precedence graphs. Prod. Oper. Manag., 1–9 (2023)
21. G.A. Kasapidis, D.C. Paraskevopoulos, Repoussis, C.D. Tarantilis, Flexible job shop scheduling problem with arbitrary precedence graphs. Prod. Oper. Manag. **30**(11), 4044–4068 (2021)
22. S. Kirkpatrick, C.D. Gelatt, M.P. Vecchi, Optimization by simulated annealing. Science **220**, 671–680 (1983)
23. D.H. Lee, Z. Cao, Q. Meng, Scheduling of two-transtainer systems for loading outbound containers in port container terminals with simulated annealing algorithm. Int. J. Prod. Econ. **107**, 115–124 (2007)

24. X. Li, X. Liang, Research on flexible job shop scheduling problem. J. Phys. Conf. Ser. **1549**, 032126 (2020)
25. X. Li, X. Guo, H. Tang, R. Wu, L. Wang, S. Pang, Z. Liu, W. Xu, X. Li, Survey of integrated flexible job shop scheduling problems. Comput. Ind. Eng. **147**, 108786 (2022)
26. S. Libo, L. Chang, S. Haibo, Z. Jun, An improved immune genetic algorithm for solving the flexible job shop scheduling problem with batch processing. Wirel. Commun. Mob. Comput. **2022**, 1–17 (2022)
27. R. Liu, R. Piplani, C. Toro, Deep reinforcement learning for dynamic scheduling of a flexible job shop. Int. J. Prod. Res. **60**(13), 4049–4069 (2022)
28. X. Long, J. Zhang, K. Zhou, T. Jin, Dynamic self-learning artificial bee colony optimization algorithm for flexible job-shop scheduling problem with job insertion. Processes **10**, 571 (2022)
29. M. Mastrolilli, L.M. Gambardella, Effective neighbourhood functions for the flexible job shop problem. J. Sched. **3**, 3–20 (2000)
30. L. Meng, Y. Ren, B. Zhang, J.-H. Li, H. Sang, MILP modeling and optimization of energy efficient distributed flexible job shop scheduling problem. IEEE Access **8**, 191191–191203 (2020)
31. B. Mihoubi, B. Bouzouia, M. Gaham, Reactive scheduling approach for solving a realistic flexible job shop scheduling problem. Int. J. Prod. Res. **59**(19), 5790–5808 (2021)
32. R. Naimi, M. Nouiri, O. Cardin, A Q-learning rescheduling approach to the flexible job shop problem combining energy and productivity objectives. Sustainability **13**(23), 13016 (2021)
33. M. Nouiri, A. Bekrar, A. Jemai, S. Niar, A.C. Ammari, An effective and distributed particle swarm optimization algorithm for flexible job-shop scheduling problem. J. Intell. Manuf. **29**, 603–615 (2018)
34. H.E. Nouri, O.B. Driss, K. Ghédira, Solving the flexible job shop problem by hybrid metaheuristics-based multiagent model. J. Ind. Eng. Int. **14**, 1–14 (2018)
35. F. Pezzella, G. Morganti, G. Ciaschetti, A genetic algorithm for the flexible job-shop scheduling problem. Comput. Oper. Res. **35**, 3202–3212 (2008)
36. M.A.F. Romero, E.A.R. García, A. Ponsich, R.A.M. Gutiérrez, A heuristic algorithm based on tabu search for the solution of flexible job shop scheduling problems with lot streaming, in *Proceedings of the Genetic and Evolutionary Computation Conference* (2018), pp. 285–292
37. O. Rossi-Doria, M. Sampels, M. Birattari et al., A comparison of the performance of different metaheuristics on the timetabling problem, in *International Conference on the Practice and Theory of Automated Timetabling* (2002), pp. 329–351
38. I. Saad, S. Hammadi, M. Benrejeb, P. Borne, Choquet integral for criteria aggregation in the flexible job-shop scheduling problems. Math. Comput. Simul. **76**, 447–462 (2008)
39. X. Shi, W. Long, Y. Li, D. Deng, Multi-population genetic algorithm with ER network for solving flexible job shop scheduling problems. PLoS ONE **15**(5), 1–23 (2020)
40. A. Van Breedam, Improvement heuristics for the vehicle routing problem based on simulated annealing. Eur. J. Oper. Res. **86**, 480–490 (1995)
41. L. Wang, J. Cai, M. Li, Z. Liu, Flexible job shop scheduling problem using an improved ant colony optimization. Sci. Program. **2017** (2017)
42. X. Wu, J. Li, X. Shen, N. Zhao, NSGA-III for solving dynamic flexible job shop scheduling problem considering deterioration effect. Intell. Manuf. **2**(1), 22–33 (2020)
43. W. Xia, Z. Wu, An effective hybrid optimization approach for multi-objective flexible job-shop scheduling problems. Comput. Ind. Eng. **48**, 409–425 (2005)
44. J. Xie, L. Gao, K. Peng, X. Li, H. Li, Review on flexible job shop scheduling. IET Collabor. Intell. Manuf. **1**(3), 67–77 (2019)
45. W. Xiong, D. Fu, A new immune multi-agent system for the flexible job shop scheduling problem. J. Intell. Manuf. **29**, 857–873 (2018)
46. M. Yazdani, M. Amiri, M. Zandieh, Flexible job-shop scheduling with parallel variable neighborhood search algorithm. Expert Syst. Appl. **37**, 678–687 (2010)
47. R. Zarrouk, I.E. Bennour, A. Jemai, A two-level particle swarm optimization algorithm for the flexible job shop scheduling problem. Swarm Intell. **13**, 145–168 (2019)

48. G. Zhang, L. Zhang, X. Song, et al., A variable neighborhood search based genetic algorithm for flexible job shop scheduling problem. Clust. Comput. **22**(Suppl 5), 11561–11572 (2019)
49. C. Zhang, Y. Rao, P. Li, An effective hybrid genetic algorithm for the job shop scheduling problem. Int. J. Adv. Manuf. Technol. **39**, 965 (2008)
50. H. Zhang, M. Gen, Multistage-based genetic algorithm for flexible job-shop scheduling problem. J. Complex. Int. **11**, 223–232 (2005)

Multimodal Freight Transport Optimization Based on Economic and Ecological Constraint

Lilia Rejeb, Abir Chaabani, Hajer Safi, and Lamjed Ben said

1 Introduction

Goods transportation plays a significant role in the present society, and with globalization growth, transport dependency is not expected to decrease in the upcoming years. The proper use of freight transport is an inherent part of the supply chain potency. For that reason, the continuous economic globalization, the growing demand for speed-to-market product delivery, and the need to manage global supply chains more effectively have led to a sustained increase in demand for multimodal freight transportation systems (MFT). A multimodal freight transportation system refers to the coordination or integration of two or more modes of transportation for delivering freight, from origin to destination, in a seamlessly linked and efficiently coordinated way. Research studies in MFT are divided into three levels: strategic, tactical, and operational. In this work, the researchers focus on the operational level, where the objective is to combine several routes, cost-effectively, served by different transportation means. The choice of modes depends on their cost, delivery time, and time flexibility (Time-flexible modes such as roads can be used whenever they are needed and schedule-based modes such as rail operate according to fixed schedules planned). It also depends on the kind of goods to be delivered.

Most of the existing research combined two modes of transport that are road and sea or road and rail. Recently interest in combining air and road to limit transit time appears. There are a minority of works that considered more than two modes [1]. In this decade, the number of multi-objective studies increased. They are in majority, bi-objective and focus on optimizing cost, time, or risk. A recent trend

L. Rejeb (✉) · A. Chaabani · H. Safi · L. Ben said
Institut Supérieur de Gestion de Tunis, SMART Lab, Université de Tunis, Tunis, Tunisia
e-mail: lilia.rejeb@isg.rnu.tn; Lamjed.bensaid@isg.rnu.tn

I. Alharbi et al. (eds.), *Advances in Computational Logistics and Supply Chain Analytics*, Unsupervised and Semi-Supervised Learning,
https://doi.org/10.1007/978-3-031-50036-7_5

was oriented to the environmental aspect, as great interest is given to Carbone's footprint, to advance sustainable multimodal freight delivery.

In this study, the researchers propose a multimodal multi-objective optimization model that considers different objectives such as time, CO_2 emissions, and cost. Several constraints are also taken into account. Besides, four transportation modes are considered, which is a massive benefit in a Multimodal Freight Transportation model. In addition, the researchers focus on one of the most critical dilemmas in today's environment which is pollution in general and CO_2 emissions in particular. When the researchers sustainably transport freight, they will have a more cost-effective, environmentally friendly, and socially inclusive freight transport and coordination [2].

There is a trend in using metaheuristic approaches in Multimodal Freight transportation. In this context, the researchers opt for the Genetic Algorithm (GA) and the Tabu Search strategy (TS) as they are well-known metaheuristics commonly used to solve NP-hard combinatorial optimization. GA has a strong global search and TS has a strong local search, high computational benefits and is less prone to be stuck in a local optimum.

This chapter is organized as follows. Section 2 summarizes the related work. Section 3 describes the model. Section 4 presents the proposed approaches. Section 5 describes the used dataset, the parameters tuning, and the comparisons of the two proposed solutions. The last section concludes the work.

2 State of the Art

In multimodal freight transport optimization, different modes of transportation play a crucial role in achieving efficient and effective logistics operations. Here is an explanation of the main modes of transportation used in multimodal freight transport optimization:

- Air Transportation: involves the use of airplanes to transport goods over long distances and across international borders. It offers fast delivery times, especially for time-sensitive and high-value goods. Air transport is suitable for perishable goods, valuable items, and goods with high demand. However, it is generally more expensive compared to other modes of transportation.
- Road Transportation: utilizes trucks and other vehicles to transport goods on roads and highways. It provides flexibility and accessibility to various locations, including remote areas. Road transport is well suited for short to medium-distance shipments and for reaching the final destination. It is often preferred for small consignments or when the delivery timeline is tight.
- Rail Transportation: involves the use of trains to transport goods over long distances on dedicated rail networks. It offers a cost-effective option for transporting bulk goods and heavy cargo. Rail transport is known for its high carrying capacity, making it suitable for large-scale shipments. It is commonly used

for transporting raw materials, commodities, and goods requiring long-distance travel.

- Sea Transportation: Sea transportation, also known as maritime transportation, utilizes ships and vessels to transport goods across oceans and seas. It is a cost-effective option for transporting large volumes of goods over long distances. Sea transport is suitable for bulky goods, such as containers, bulk cargo, and commodities. It is commonly used for international trade and global supply chains.

In multimodal freight transport optimization, these different modes of transportation are combined strategically to leverage their respective strengths and mitigate limitations. By utilizing a combination of air, road, rail, and sea transportation, logistics providers can optimize routes, minimize costs, improve delivery times, and reduce environmental impacts, ultimately achieving efficient and sustainable freight transportation operations. The researchers detail, in the following, the state-of-the-art research that implemented different types of MFT path-planning problems to improve the quality of the routes and display the performance of several approaches.

Zufferey and Verma [3] suggested planned and routed rail-truck shipments from a set of suppliers to a set of clients. The delivered quantities follow a multimodal route of truck-rail-truck. Three requirements were met: each request must be provided by the due date, the demand must be satisfied, and the train's capacity cannot be surpassed. A weighted average of the cost and the risk serves as the objective function to minimize. The Tabu Search (TS) algorithm was used to solve this issue.

To evaluate the possibility of increasing the absorption rate by a multimodal transport service, Pereira et al. [4] created a parameterized model for the optimization of a multimodal transportation network (Road-sea). A novel solution was considered combining the differential evolution strategy with constructive heuristic.

Lei et al. [5] proposed a mathematical model for multimodal transportation scheme decision optimization based on transportation costs, time, and risks. To find solutions, they introduced the PSACO algorithm which combines the Particle Swarm (PS) and Ant Colony Optimization (ACO), absorbing their strengths and overcoming their weaknesses. Its advantages are complementary and superior to those of the ant colony algorithm in terms of time efficiency and the particle swarm algorithm in terms of accuracy and efficiency.

To examine the intermodal transportation planning problem in a stochastic and dynamic context, a hybrid simulation and optimization approach was presented by Hrusovsky et al. [6]. In this work, the authors optimized the transportation plan by boosting reliability in terms of journey time uncertainty. They propose greener transportation solutions for the problem under investigation. A case study from real life was used to validate the proposed approach.

Sun et al. [7] proposed a nonlinear model to solve the real-world problem considering road traffic congestion and uncertainty in the capacity of rail services thoroughly. This model considered CO_2 by injecting its costs in a bi-objective viewpoint and determined if the method is better for designing environmentally friendly routes. The proposed nonlinear model was linearized to be resolved by an exact approach.

To reduce overall costs in an intermodal sea-rail network, Zhao et al. [8] formulated the Stochastic Intermodal Service Network Design (SISND) problem as a two-stage chance-constrained programming model with stochastic transit time, transfer time, and container demand. It was suggested that the SISND problem could be solved under capacity and on-time delivery limitations with predefined confidence levels, using a hybrid heuristic algorithm including the SAA technique and ACO algorithm. To show the feasibility of the suggested concept, a numerical example was run on an intermodal sea-rail network from China to Singapore.

Heggen et al. [9] gave a recommendation for an integrated multimodal routing heuristic that supports the integration of decisions about truck routing and the choice of long-haul multimodal services. The approach aimed at reducing the distance and maximizing the long-haul capacity utilization. In this way, it makes the modal shift easier and leads to a more sustainable transport system. This work focused on a two-region truck-rail network. However, the proposed approach can deal with other transport modes and multiple regions.

Sun et al. [10] looked into the multimodal hazardous materials road-rail routing issue, which is a hot topic in the realm of transportation planning. They first employed a hub-and-spoke network to depict the hazardous materials road-rail multimodal transportation network; second, they considered the uncertainty of the risk parameter, i.e., the population exposures; and third, the sustainability of the hazardous materials transportation was enhanced by formulating a carbon dioxide emission constraint to lower greenhouse gas emissions. The problem was formulated as a bi-objective fuzzy mixed integer nonlinear programming model. The authors then developed a three-stage exact solution strategy, where they combined fuzzy credibility chance constraints, linearization technique, and the normalized weighting method.

Zhang et al. [11] investigated the low-carbon path optimization problem under uncertainty. They proposed a Hybrid Robust stochastic optimization (HRSO) model considering the transportation cost, time cost, and carbon emission cost. To solve this problem, they used a Catastrophic Adaptive Genetic Algorithm (CA-GA) based on Monte Carlo sampling. It was tested on multiple networks.

Ge [12] proposed a bi-objective model to determine the most reasonable transportation path. It considered the minimum cost and time as goals to optimize. It used an ant colony algorithm to predict the path.

Terminals are the foundation of each intermodal freight system. Oudani [13] studied the Intermodal Terminal Location Problem on incomplete networks. He modeled this problem as a mixed integer linear program. He developed a simulated annealing algorithm to tackle medium and large instances. The obtained solutions using simulated annealing were competitive to the exact solutions found by CPLEX solver for small and medium instances.

Sun et al. [14] proposed fuzzy multi-objective nonlinear optimization model for the road-rail intermodal routing problem. They aimed to minimize the total costs and carbon dioxide emissions of the routes. They took in account uncertainties of travel times and carbon dioxide emission factors of road services and capacities of rail services. They solve the model by designing an interactive fuzzy programming approach with the Bounded Objective Function method.

Lu et al. [15] proposed a multi-objective chance-constrained model considering three objectives: time, cost, and CO_2 emission and two modes: railways and highways. It considers also the uncertainties of freight volumes, expenses, time, and carbon emissions. They used NSGA-III to solve the model.

Li et al. [16] established a hybrid robust stochastic optimization model of container multimodal transport considering costs and CO_2 as objectives and rail, highways and water as modes. They considered the uncertainty of demand and the random price of carbon trading. To solve the problem under double uncertainty they designed a fireworks algorithm with an attractive force search operator.

Okeyere et al. [17] formulated an optimal Sustainable Multimodal Freight Transport and Logistics System (SMFTLS) framework that optimizes distance, time, and CO_2 emissions. They considered three transport modes (road, rail, and ship) to solve the formulated problem.

In the work of Yang et al. [18] a multi-objective optimization model was established with transportation distance, transportation time, and carbon emission as transportation objectives. They designed an improved fuzzy adaptive genetic algorithm (FAGA) to solve the problem.

As presented previously, the majority of the state-of-the-art focused on rail and road transportation or road and sea (See Table 1) and rarely on three modes. Moreover, the majority of works considered a single objective or two objectives. They minimized cost with time or risk or CO_2. There are few works that try to combine three objectives together. These findings are confirmed by Archetti et al. [1]. They proposed a review of multimodal freight transportation problems modeled through optimization. They tackled the problem from a strategic, tactical, and operational level. They concluded that there is a lack of research in the operational level where we find complex problems. The existing works considered mostly two or three modes (rail, sea, and road). A recent interest was given to planes due to the reduced delivery time requested. Archetti et al. [1] noticed, also, that the number of works based on multi objective optimization is increasing and that the latter is better to model the trade-off between cost and delivery time. They certified that it is interesting to consider also CO_2 emissions and risk.

Motivated by these issues, this work presents a model that takes into consideration all four modes (trucks, trains, ships, and planes) and three different objectives that are the overall transportation cost, time, and CO_2 emissions. The main target is to propose a model closer to the reality and to obtain an economic, ecological, and green freight transportation that guarantees the client satisfaction in terms of delivery time.

To find a solution for the multimodal freight transportation problem, different approaches, varying from exact to approximate ones, were used in the literature (see Table 1). Using different kinds of metaheuristics is an actual trend. Metaheuristics are recommended since they are capable of finding satisfying solutions in reasonable time even for large instances [1]. The researchers adopt two different metaheuristics

Table 1 Synthesis of works on multi-model transportation

Paper	Modes	Nature	Technique	Objectives
Zufferey and Verma [3]	Trucks, trains	Deterministic	Tabu search	Weighted summation of the cost and the risk
Pereira et al. [4]	Road, Sea	Deterministic	Hybrid Differential Evolution and a heuristic	Costs, Fuel consumption
Lei et al. [5]	Not specified	Deterministic	Ant colony and particle swarm optimization algorithm	Costs time and risks
Hrusovsky et al. [6]	Road, rail and barge	Stochastic (Transportation time)	Hybrid simulation and optimization approach	Costs, delay penalties and emissions
Sun et al. [7]	Road, rail	Stochastic (Transportation time and capacity)	Branch-and-bound algorithm	Costs, delay penalties and emissions
Zhao et al. [8]	Rail, sea	Stochastic (Container Demand and Transportation time)	Ant colony optimization	Cost and delay penalties
Heggen et al. [9]	Road, Rail	Deterministic	Large neighborhood search (sequential and integrated approach)	Cost
Sun et al. [10]	Road, Rail	Stochastic (risk parameter)	Commercial solver: The normalized weighting method	Cost, early delivery penalties and risk exposure
Ge [12]	Not speci-fied	Deterministic	Ant colony algorithm	Cost and time
Oudani [13]	Road and rail	Deterministic	Simulated Annealing	Cost and CO_2
Zhang et al. [11]	Road, rail	Stochastic (Time and Demand)	Catastrophic adaptive Genetic Algorithm (CA-GA) based on Monte Carlo sampling	Cost time and CO_2
Lu et al. [15]	Rail, highway	Stochastic	NSGA-III	Cost and CO_2
Li and Sun [16]	Rail, Sea and highway	Stochastic	fireworks algorithm	Cost time and CO_2
Sun et al. [14]	Rail and Road	stochastic	interactive fuzzy programming approach	Cost and CO_2
Okeyere et al. [17]	Road, water and train	Deterministic	Genetic Algorithm	Cost and CO_2
Yang et al. [18]	Road and train	Deterministic, Stochastic	fuzzy adaptive genetic algorithm (FAGA)	Distance, time and CO_2

that are the TS which is based on local search and the GA which is a population-based algorithm.

3 MFT Problem Description

The problem is modeled on a directed graph where nodes represent the locations such as ports, airports, and warehouses and the arcs represent the transportation services between locations and the modes used depending on the locations themselves. If the concerned locations are two airports, for example, the mode is air. Each commodity must be shipped from its origin to its final destination by the order date and deadline. Our considered problem has to output an ordered list of the given routes that enables each good to get to its destination using different modes while respecting several constraints. It is considered then as a multi-objective, multimodal, multi-commodity flow problem. In the state of the art, only up to three modes were used (sea, truck, and rail). To be closer to reality, we consider also air transport. The kind of mode depends also on the good to be delivered. We consider in our case goods that are transported in containers. In our problem, we deal with three different objectives whereas the majority of the proposed works deal only with two of them.

Our goal is to simultaneously reduce the overall cost of traveling, the travel time, and the travel CO_2 emissions. Figure 1 describes the considered MFT problem.

In the following, we describe the hypotheses taken into consideration with the resolution part, since there are too many immeasurable factors that can influence the delivery process and outcomes in real-world transportation situations. The primary presumptions are as follows:

- The delivery method is deterministic; therefore, neither delivery time nor cost will be affected by random events.
- There is no need for specialized containers (refrigerated, thermostatic, etc.) to transport goods; they can be transported in regular containers.
- The volume of the good is only constrained by the container, and the volume of all goods can be divided (A bin packing issue was not necessarily taken into account).
- The model solely assesses the principal carriageways. The first and last miles, between the end user and the origin or destination shipping point, are not taken into account (From one warehouse to the next).
- Between each pair of ports, a single transit tool is provided. For instance, a direct journey by ship, railroad, or truck between two airports in different cities is impractical. However, we can travel directly between them via flying.
- The three most crucial components of the total cost—warehouse costs, transportation costs, and goods tariffs—are the only ones included in the model.
- A day is the smallest unit of time, and a route can only have one trip in a day.
- The transportation cost is computed only regarding the trip cost, the tax cost, and the warehouse cost.

Fig. 1 Diagram illustrating the MFT

Our MFT path-planning problem is a multi-objective optimization problem inspired from the work of [7, 19–21]. The decision variables, dimensions are presented in the next subsections.

3.1 Dimensions

The researchers consider five dimensions in our work that are:

- **Start Port (i):** Identifies the port where a direct transport route begins. The number of ports in all the data is equal to the dimension length.
- **End Port (j)**: Indicates a direct transport route's final port. The number of ports in all the data is equal to the dimension length.
- **Time (t):** States the time when a direct conveyance will depart. The length of the dimension is the sum of the days between the oldest order date and the latest delivery deadline date for all items in the data.
- **Goods (k):** Indicates the goods to be transported. The dimension length equals the total number of goods in the data.
- **Modes (l, p):** Identifies the types of modes used to transport the different freight.

3.2 Decision Variables

Decision variables are:

- **Matrix of Decision Variables (X):** where an element $X_{i,j,t,l}$ indicates whether the route from node i to j is used by the mode l at time t. The matrix has four dimensions. Each one standing for the start port, finish port, time, and mode. Each component of the matrix is a binary variable.
- **Matrix of goods (W):** where an element $W_{i,j,t,k}$ indicates whether the goods k are transported on the route from node i to j at time t. W serves as the decision variable matrix's foundation. It is a four-dimensional matrix with start port, end port, goods and time as dimensions. Each component of the matrix is a binary variable, representing whether a route is used for a specific freight.
- **Matrix of Container Numbers (Y):** The number of containers required to load all products moving concurrently from port i to port j using mode l at time t is represented by $Y_{i,j,t,l}$. Y is a matrix that serves as the decision variable matrix's foundation. Start port, end port, mode, and time represent different aspects of the matrix's four dimensions.
- **Matrix of Decision Variables (S):** where an element $S_{i,k,l,p}$ indicates whether commodities k are transshipped from mode l to mode p in port I. The matrix has four dimensions, each one standing for the current port, time, goods, and modes, respectively.

3.3 Parameters

In this part of the work, parameters are depicted as follows:

- **Cost per container (C):** The total cost of shipping a container from port i to port j at time t using mode l is shown in the matrix as an element $C_{i,j,t,l}$. This total cost includes handling fees, bunker/fuel fees, paperwork fees, equipment fees, and other fees. The cost component for an impractical route will be set to big M (a very large value), rendering the option impossible. C is a four-dimensional parameter matrix with the start port, finish port, modes, and time values assigned to each dimension.
- **Fixed Cost for Route (FC):** where $FC_{i,j,t,l}$ reflects the fixed transportation cost to travel from port i to port j at time t using mode l. The cost element will also be set to big M for routes that are impractical. FC is a four-dimensional parameter matrix with the start port, finish port, and time values assigned to each dimension, regardless of the quantity or volume of the commodities.
- **Warehouse Cost (wh):** Is an array with one dimension which is the port i. It defines the warehousing cost per cubic meter per day. Warehouse costs are expected to be big M for ports without a warehouse function (such as airports, railroad stations, etc.).

- **Tax Percentage (tax):** The tax percentage tax k on products k is what the country of destination has levied. The tax rate for commodities that are only transported domestically will be fixed at 0. It is a single-dimensional array.
- **Transit Duty (td):** The transit duty (tax levied on products passing through a nation) percentage for goods traveling from port i to port j is represented by $td_{i,j}$. Transit duty percentage is set to zero if ports i and j are in the same nation. For the sake of simplicity, the transit fee is fixed at the same rate for all commodities (It could be easily extended). It is a two-dimensional matrix with the start port and finish port being represented by each dimension.
- **Transit hours (Tt) :** The matrix's entry $Tt_{i,j,t,l}$ indicates the transportation time on route between port i and port j using mode l on t time. It is a four-dimensional parameter matrix with the start port, finish port, modes, and time values assigned to each dimension.
- **Transshipment hours (Ts):** $Ts_{i,k,l,p}$ represents the time for transshipping goods k from mode l to mode p at the node i. It is a four-dimensional parameter matrix with the start port, modes, and goods values assigned to each dimension.
- **Transportation Time (T):** The matrix's entry $T_{i,j,t,l}$ shows how long it takes to go from port i to port j at time t using mode l. This total time includes the time required for customs clearance, handling, transit, and additional time. The time element will be configured to be big M for routes that are unfeasible. T is a four-dimensional parameter matrix with the start port, finish port, modes and time values assigned to each dimension.
- **Container Volume (ctnV):** The transported container volume on the route from port i to port j is represented by $ctnV_{i,j}$. It is two-dimensional matrix with the start port and finish port being represented by each dimension.
- **Goods Volume (V):** V_k stands for the quantity of items k. It is a one-dimensional array of goods.
- **Goods Value (val):** The Value of the commodities k is represented by val_k. It is a one-dimensional array of goods.
- **Order Date (ord):** The order date for item k is denoted by the symbol ord_k. It is a one-dimensional array.
- **Deadline Date (ddl):** ddl_k stands for the target delivery date of the specified products in a one-dimensional array with goods as dimension.
- **Origin Port (OP):** The port where goods k originate represented by OP_k. It is a one-dimensional array of goods in dimensions.
- **Destination Port (DP):** The port represented by DP_k is the location of the items in k. It is a one-dimensional array of goods in dimensions
- **Transportation CO_2 emission (EC):** CO_2 emission resulting from transporting k from port i to port j using mode l is represented by $EC_{i,j,k,l}$. The CO_2 emission depends highly on the mean used, the distance, and the weight of goods. EC is a four-dimensional matrix with start and end port, goods and modes as dimensions.
- **Transshipment CO_2 emission (ec):** CO_2 emission caused by loading goods k from mode l to p in node i is represented by $ec_{i,k,l,p}$; it is a four-dimensional matrix with start a port, goods and modes as dimensions.

The current multi-objective optimization problem's objective functions and constraints are defined as follows.

3.4 Objectives

1. The first objective minimizes the total cost of the overall transportation which includes the trip cost, the tax cost, and the warehouse cost.

$$min_{x,y,w} OT = transportation\ cost + tax\ cost + warehouse\ cost. \quad (1)$$

- Transportation cost:

$$\sum_{l}\sum_{t}\sum_{j}\sum_{i} Y_{i,j,t,l} * C_{i,j,t,l} + X_{i,j,t,l} * FC_{i,j,t,l} \quad (2)$$

- Tax cost:

$$\sum_{k} tax_{k} * val_{k} + \sum_{i}\sum_{j}\left(\sum_{t}\sum_{k} W_{i,j,t,k} * val_{k}\right) * td_{i,j} \quad (3)$$

- Warehouse cost:

$$\sum_{i}\left[\sum_{k}\left(\sum_{j}\sum_{t} t * W_{i,j,t,k}\right) * V_{k}\right] wh_{i}$$
$$-\sum_{j}\sum_{k}\left[\sum_{i}\sum_{t}\sum_{l}\left(t + T_{i,j,t,l}\right) * W_{i,j,t,k}\right] * V_{k} * wh_{j} \quad (4)$$

2. The second objective is to minimize the overall travel time from the origin port to the destination one. It is composed of the transportation time on route and the transshipment time in nodes. It is presented by the Eq. (5).

$$MinTo = \sum_{i}\sum_{j}\sum_{t}\sum_{l} Tt_{i,j,t,l} * X_{i,j,t,l} + \sum_{i}\sum_{k}\sum_{l}\sum_{p} Ts_{i,k,l,p} * S_{i,k,l,p} \quad (5)$$

3. The third objective is to minimize the overall CO_2 emission. The minimization of the overall CO_2 emission is the minimization of the CO_2 emission on route

and CO_2 emission of transshipment in nodes Eq. (6).

$$MinCo = \sum_i \sum_j \sum_l EC_{i,j,k,l} * X_{i,j,t,l} + \sum_i \sum_k \sum_l \sum_p ec_{i,k,l,p} * S_{i,k,l,p} \quad (6)$$

The three objectives are not conflicting; therefore, we can sum them in a mono-objective function:

$$MinComp = MinOT + CMinTo + CMinCo \quad (7)$$

Where $CMinTo$ and $CMinCo$ are, repsectively, the transformations into cost of the $MinTo$ (time) and $MinCo$ (CO_2 emission).

There is no preference between the functions. They are all with the same importance.

3.5 Constraints

1. Each good k must be transported from its origin to a different node and then transported to its destination.

$$\sum_t \sum_j W_{OP_k,j,t,k} = 1, , \forall k \sum_t \sum_i W_{i,DP_k,t,k} = 1, \forall k \quad (8)$$

2. For each good k, neither its origin nor its destination could be used for shipping (No shipping OUT of destination or shipping TO origin).

$$\sum_t \sum_j W_{OP_k,j,t,k} = 0, , \forall k \sum_t \sum_i W_{i,DP_k,t,k} = 0, \forall k \quad (9)$$

3. Ship-in times and ship-out times must match for each good k at transition point j (neither origin nor destination).

$$\sum_t \sum_i W_{i,j,t,k} = \sum_t \sum_i W_{j,i,t,k} \forall k, j \mathbf{s.t}, j \neq OP_k, j \neq DP_k \quad (10)$$

4. At most one transfer of each good k into or out of a port is permitted.

$$\sum_t \sum_j W_{i,j,t,k} \leq 1, \forall k, i \quad (11)$$

$$\sum_t \sum_i W_{i,j,t,k} \leq 1, \forall k, i \quad (12)$$

5. Ship-out time needs to come after ship-in time for each good k at the transition port j. Ship-out time for commodities k at their origin port should come after the order date since there is no ship-in time as stated in Eqs. (13) and (14).

$$\sum_j \sum_t t * W_{i,j,t,k,l} - \sum_j \sum_t (t * W_{j,i,t,k} + T_{j,i,t,l} * X_{j,i,t,l}) \geq 0, \forall k, i\mathbf{s.t}, i \neq OP_k, i \neq DP_k \tag{13}$$

$$\sum_j \sum_t t * W_{OP_k,j,t,k} - \sum_j \sum_t (t * W_{j,OP_k,t,k} + T_{j,OP_k,t,l} * X_{j,OP_k,t,l}) \geq ord_k, \forall k \tag{14}$$

6. The total volume of containers at each route at time t should be greater than the total volume of merchandise.

$$Y_{i,j,t,l} \geq \frac{\sum_k W_{i,j,t,k} * V_k}{ctnV_{i,j}}, \forall i, j, t \tag{15}$$

7. This constraint verifies if a route is being used at time t. Where $W_{i,j,t,k}$ is a binary variable, and $X_{i,j,t,l}$ for all items k at i,j,t must be greater than 0 if a route is employed. By multiplying a small number, we can reduce the range to [0,1].

$$X_{i,j,t,l} \geq \sum_k W_{i,j,t,k} * 10^{-5}, \forall i, j, t \tag{16}$$

8. Each item k must be delivered by the deadline of the delivery date to its destination port.

$$\sum_i \sum_t (t * W_{i,DP_k,t,k} + T_{i,DP_k,t,l} * X_{i,DP_k,t,l}) \leq ddl_k, \forall k \tag{17}$$

9. The total transportation time should not exceed the transit period of goods.

$$\sum_i \sum_j \sum_t \sum_l Tt_{i,j,t,l} * X_{i,j,t,l} + \sum_i \sum_k \sum_t \sum_l \sum_p Ts_{i,k,l,p} * S_{i,k,l,p} \leq \sum_i \sum_j \sum_t \sum_l T_{i,j,t,l} \tag{18}$$

10. There is no transshipping at the origin node and destination node.

$$S_{\mathrm{i,k,l,p}} = 0, \forall i \in \{OP, DP\}, \forall k, l, p \tag{19}$$

11. Only one kind of transport mode can be selected between determined nodes i and j.

$$\sum_{l} X_{\mathrm{i,j,t,l}} = 1, \forall i, j, l \tag{20}$$

12. The decision-making variables are binary, taking integer values only equal to 0 or 1.

$$X_{\mathrm{i,j,t,l}}, W_{\mathrm{i,j,t,k}}, , S_{\mathrm{i,k,l,p}} \in \{0, 1\} \forall i, j, l, k, p \tag{21}$$

4 Proposed Solution Methods

To solve the MFT problem, we design two solution approaches which are the Tabu search and Genetic Algorithm. Our choice is justified by the following reasons: Firstly, the genetic algorithm [22] is characterized by a unique parallelism and a probabilistic nature, which are appropriate for global optimization and convenient for a large set of space solutions. GA is chosen over other optimizers because it handles several solutions at one generation leading them to a higher exploration rate. This property may be suitable for multimodal transport problems. In addition, the Tabu Search metaheuristic has proven its efficiency in solving several NP-hard combinatorial problems due mainly to the strong local search procedure. TS negotiates a local minimum while keeping track of previous searches in its memory. It is an intelligent search technique that hierarchically canalizes one or more local search procedures to quickly search the global optimality and uses a memory function to avoid being trapped at a local minimum [23].

Taking into consideration these arguments, we are motivated through this work to exploit the main features of both metaheuristics in solving the MFT problem. We describe, in the following subsections, the step-by-step details of the developed methods.

4.1 Tabu Search-Based Solution

The algorithm description of the adopted TS strategy is described through Algorithm 1.

Algorithm 1 TS based MFT multi-objective path-planning optimization

```
Input: The routes Graph G =(V, E), the origin O and destination D nodes, the MFT's parameters, the maximum number of function evaluation fe, tabu list, tabu list size, tabu tenure, Candidate list.
Output: best route.
Step 1 generate initial solution S0 randomly
tabuList ← []
currentSolution ← S0
bestSolution ← S0
while i < fe do
        bestCandidate ← null
        Step2: Generate the set of neighborhood solutions of the current solution feasible routes.

    for candidate ∈ currentSolution.getNeighbourhood do
        Step3: check if the candidate route in the tabu list

        if ¬tabuList.contains(candidate) then
            Step4: choose best admissible candidate in terms of route, CO2, time and cost

            if (fitness(candidate) < fitness(bestCandidate)) then
                bestCandidate ← candidate
            end if
        end if
    end for
    Step5: Find the best solution from the candidate list
    currentSolution ← bestCandidate

    if fitness(bestCandidate) > fitness(bestSolution) then
        bestSolution ← bestCandidate
    end if
    Step6: Update TL and aspiration condition
    tabuList.push(bestCandidate)

    if tabuList.size > tabuTenure then
        tabuList.removeFirst()
    end if
end while
return bestSolution
```

In the initialization step, we begin by generating an initial solution S0 using a random function and set it as the current solution and best solution (Step 1). In the second step, we generate a set of neighbors of the current solution which represents feasible routes (Step 2). The generation is done using a neighborhood function. In our case, the neighbor consists of swapping the position of two nodes within the route while respecting the constraints (The neighborhood set was generated by a two-point exchange method: two random elements of the solution are swapped concerning the constraints). After generating the neighbors, solutions are evaluated using the aggregate objective functions. If a candidate's solution fitness is better than the fitness of the best solution then an aspiration criterion is met and the solution that satisfies the aspiration criteria is then set as the best one. At this stage, the Tabu list is updated. If the aspiration criterion is not met then the best and current solution which is not a Tabu solution is selected from the neighborhood solutions and set as the current solution. The Tabu list is updated by adding a new solution to it (Step 4). This process will go on until a termination criterion is fixed by several function evaluation parameters. Then the best solution that represents the optimal route is outputted.

4.2 Genetic Algorithm-Based Solution

The description of the adopted GA strategy is described through the steps presented in Algorithm 2.

After setting all its parameters, the GA starts with generating an initial population P of N random feasible paths (solutions) from the starting point (O) to the destination (D) (Step 1). These individuals are encoded as chromosomes. Each one is evaluated based on the objective functions and constraints (Step 2) and then ranked by that value. The rank selection operator orders the individuals according to their fitness in order to select the best two parents to proceed with the crossover (Step 3,4). At this point, the single-point crossover is used to create two new offspring from the selected mating pool (Step 5). According to the mutation probability, single-point mutation can occur slightly, changing the produced offspring and giving birth to newer one (Step 6). The offspring are evaluated according to the constraints (Step 7). The old population is replaced by the newer one (Step 8) and the process is

Algorithm 2 GA based MFT multi-objective path-planning optimization

Inputs: The routes Graph **G =(V, E)**, the origin **O** and destination **D** nodes, the MFT's parameters, the maximum number of function evaluation fe, the mutation probability **mp**, the crossover probability **cp**, the population size N.
Output: Best global solution.
Step 1: Generate the initial population P_0 of N random individuals from **O** to **D**
while $i < fe$ **do**
    for i=1..N **do**
        Step 2: Evaluate the fitness of each individual chromosome
        Step 3: Assign a rank to each chromosome based on its fitness
        Step 4: Select two random parents $parent_1$ and $parent_2$ from the P_t using the rank selection
        Step 5: Produce two offsprings O_1 and O_2 using single-point crossover
        if $randValue \leq$ **mp then**
            Step 6: Mutate O_1 and O_2
        end if
        Step 7: Evaluate constraints for each offspring O_f
    end for
    Step 8: Replace the new population
end while
return best route.

repeated until we reach the stopping criteria, which is fixed by a number of function evaluations. The GA then renders the best route (Step 9).

Other details regarding the development of the GA such as the coding scheme, fitness function, genetic operator, and operating parameters are detailed in the following subsections.

4.3 Chromosome Encoding

In our GA, the chromosome is represented in a "Permutation form." A position gene represents the location ID, while the odd position gene indicates the transport mode code. The ends of the head and tail have determined values, marking the start and end locations. A positive integer is used to represent the location ID, which is derived from the multimodal transport network topology. The code for the transportation mode utilizes a negative number to separate it from the location, thus −1, −2, −3, and − 4 stand for truck, train, ship, and plane, respectively [17]. The path from site 1 to position 2 is depicted in Fig. 2. In between, the route passes through two locations with the IDs 5 and 6, respectively. The path goes from location 1 to location 5 using a truck, then from location 5 to location 6 using a ship, and finally from location 6 to location 2 using a truck. Figure 3 presents another example of the chromosome encoding. It shows the path from site 7 to site 8 crossing sites 5, 6, 2, and 3.

4.4 Selection Operator

In this study, the rank selection method is employed to identify a limited number of individuals that will proceed to the next generation of the population. By ranking the population, rank selection evaluates the fitness of each chromosome. The lowest fitness level is 1 and the highest is N. It produces steady convergence

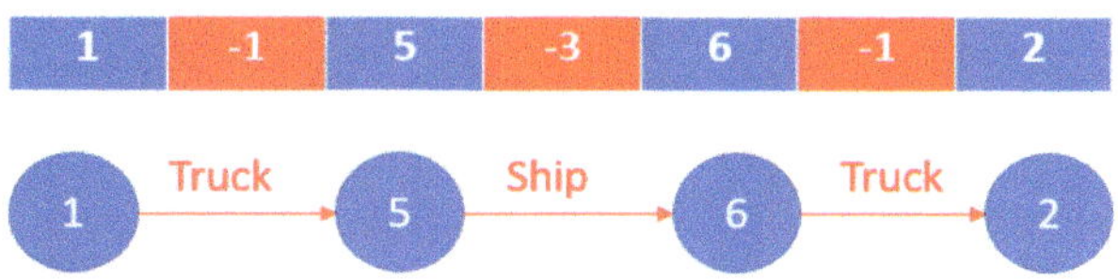

Fig. 2 Chromosome encoding example 1

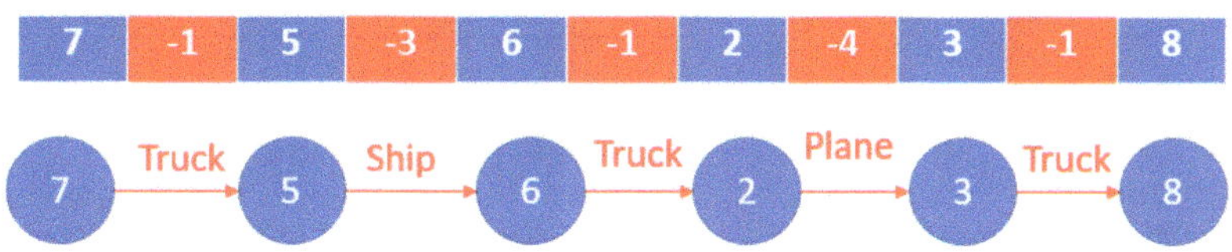

Fig. 3 Chromosome encoding example 2

while avoiding rapid convergence. The main benefit of rank-based selection is that it controls the pressure of selection, which improves the algorithm convergence [24]. By "convergence," we indicate the situation of the search process stagnation brought on by similarities among population members. When the fitness variance is small, selection pressure is maintained as well. It protects diversity, which results in a fruitful search. This selection can be done in many ways, but we suggest the following one: choose two chromosomes at random. The individual who receives the highest fitness evaluation becomes the parent and we continue looking for a second parent.

4.5 Variation Operators

Crossover and mutation processes offer search capabilities and speed up convergence. To shorten the computation time per generation, the researchers adopted the steady-state model [25] with single-point crossover and swap mutation.

In the single-point crossover adopted, for the selection of intersections, all locations excluding the start and the ending were chosen. For the code string representing the solution to the problem, the position number starts from 0, and the size indicates the length of the chromosome. It is an odd number, and the intersection position is chosen at random in sizes 2, 4, etc.

In the combination process, if the intersection of chromosome A and the intersection of the chromosome B are exactly at the same places (odd place for location ID and pair one for mode), then the direct combination can correctly generate two new paths; otherwise, another combination must be made to respect the constraints [26].

As shown in Fig. 4, the crossover point is made at a location ID (odd place) for both feasible solutions. The offspring are feasible after being generated with respect

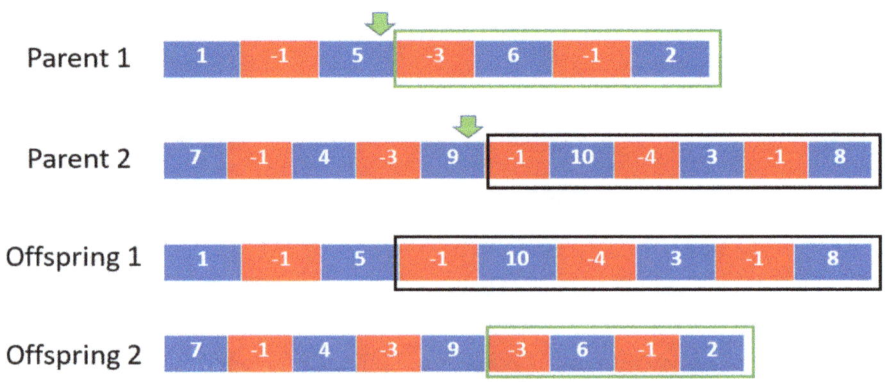

Fig. 4 Single-point crossover

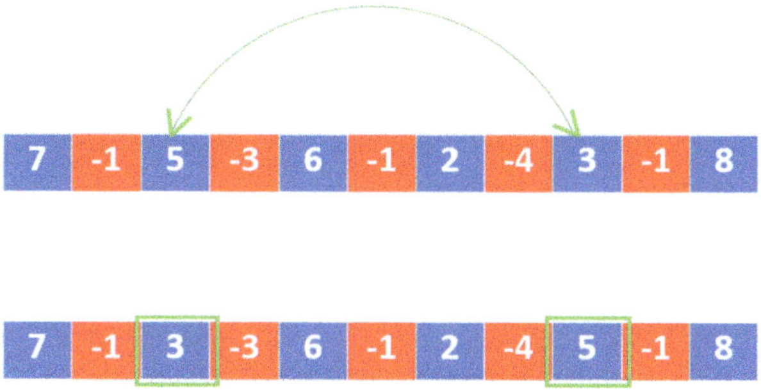

Fig. 5 Swap Mutation

to certain constraints such as no returning back to the same location, the source is different from the destination, and only one mode is used between two different locations.

Following the crossover, the "Swap Mutation" operator is performed (Fig. 5), only if the randomly generated value is lower or equal to the mutation probability (i.e., random value $\leq$ mp). For the "Swap Mutation," we pick two random positions in the chromosome and swap the values there. In permutation-based encodings, this operator is common [27]. The two selected positions must be from the same place (odd or pair).

5 Experimentation and Results

To evaluate the two used solution approaches, the researchers present firstly the dataset. Then they present the parameters tuning of each method used in the comparative results' subsection.

5.1 *Data Source Description*

The data was provided by Ken Huang on GitHub.[1] The file contains two different sheets, "Order Information" and "Route Information." The first one includes data about the order and shipping of eight types of freight: honey, furniture, paper plates, durian, cigarettes, apples, and pharmaceutical drugs. Each shipment has an order number, a weight and volume, and its source and destination.

Ken Huang provides, in the second sheet, 50 direct routes linking several ports in Wuxi, Singapore, Malaysia, and Shanghai. Every route has a unique transportation mode, cost, travel time, and weekly timetable for transportation. Each city/country has a warehouse where items can be stored for a while to match specific transit

[1] https://github.com/hzjken/multimodal-transportation-optimization.

Table 2 Freights information

Freight	Ship From	Ship To	Order Date	Required delivery date	Weight (Kg)	Journey type
Honey	Singapore Warehouse	Wuxi Warehouse	01-Feb-18	25-Feb-18	21,000	International
Furniture	Malaysia Warehouse	Shanghai Warehouse	02-Feb-18	23-Feb-18	20,000	International
Paper Plates	Singapore Warehouse	Shanghai Warehouse	03-Feb-18	23-Feb-18	20,000	International
Pharmaceutical drugs	Singapore Warehouse	Shanghai Warehouse	04-Feb-18	24-Feb-18	20	International
Cigarette	Wuxi Warehouse	Malaysia Warehouse	05-Feb-18	22-Feb-18	5000	International
Apple	Shanghai Warehouse	Singapore Warehouse	06-Feb-18	23-Feb-18	25,000	International
Durian	Malaysia Warehouse	Singapore Warehouse	07-Feb-18	24-Feb-18	10,000	International
Furniture 2	Wuxi Warehouse	Shanghai Warehouse	08-Feb-18	25-Feb-18	20,000	Domestic

timetables or to wait for other goods to be carried together. This file considers only transportation costs and time.

These data were augmented by the distance between ports and the emission factor or coefficient (specific to each mode and measured in kilograms (kg) of CO_2 produced per megawatt-hour (MWh) of energy generated from a specific source of fossil fuel). According to the commodities' weight (W), the distance between points (D), and the emissions factor (EF) we can compute the CO_2 emissions.

For each freight, we calculated its CO_2 emissions for each route from one point to another in the data. The calculations were carried out using a CO_2 emission calculator named "CarbonCare",[2] which is a certified "online worldwide CO_2 emissions calculator" for all modes of transport (road, rail, air, sea, and inland waterways), including emissions from cargo handling and cold storage.

Table 2 shows the source and destination of each shipment, along with their order date and the required arrival date, which is the deadline that we must respect. The second furniture freight is the only freight shipped domestically (it does not cross borders).

Each port has working days. There are days when, for example, airports or seaports do not accept incoming shipments. However, trucks are available all days of the week. Moreover, transshipment, which is the handling of goods when transitioning from one mode to another in a specific location, takes from a day up to

[2] https://www.carboncare.org/en/co2-emissions-calculator.html.

Table 3 TS parameters values

Parameters	Values
Number of function evaluation (termination criteria)	1500
	2000
Tabu list size	20
	30
	40
Aspiration criteria	Yes
	No

2 days according to the mode in question. For example, handling seaports can take 2 days while handling rails takes, only, 1 day.

5.2 Parameters Exploration

In this subsection, we define the different input parameters used by our algorithms. We investigate these parameters tuning to fix their optimal configuration. For each parameter tuning, the other parameters remain fixed.

5.2.1 Tabu Search Parameters Exploration

The researchers start with the implementation of the first algorithm TS. Table 3 presents the different parameters and their chosen values. We should note that for each combination set, we turn the algorithm 30 times to reach a reliable average performance.

Number of Function Evaluation To evaluate the effect of the number of function evaluations on the TS algorithm, the researchers run the program using two different evaluation number values: 1500 and 2000. The tabu list size is fixed to 30 and the aspiration criteria is considered. Table 4 presents the cost, time, and CO_2 results of the entire journey. We consider the eight freights along with the computational time of the algorithm. We can deduce that the solution improved for each objective going from 458439.0 to 402643.0 for the cost, from 360963.54 to 309387.36 for the CO_2 emissions, and from 69 to 63 days, for the overall time. However, the computational time increased from 2320 (s) to 2967 (s) because the researchers iterate more to meet the evaluation numbers. We can conclude then that the greater the evaluation number is, the better the results are. With a big evaluation number, the TS can evaluate more solutions better in the diversified search space.

Tabu List size From Table 5, we can conclude that the smallest size of the list produced the worst solution. The best solution was produced with a tabu list size of 30, where we can see that the cost improved from 491527.0 to 402643.0, the CO_2

Table 4 TS: Comparison of the objective functions according to the function evaluation number

Evaluation number	1500	2000
Total cost	458439.0	402643.0
Total CO_2	360963.54	309387.36
Total time	69	63
Computational time (s)	2320	2967

Table 5 TS: Comparison of the objective functions according to the tabu list size

Tabu list size	15	30	45
Total cost	491527.0	402643.0	448711.0
Total CO_2	335994.8	309387.36	321039.72
Total time	72	63	67
Computational time (s)	2975	2967	2960

Table 6 TS: Comparison of the objective functions according to the aspiration criteria

Aspiration criteria	Yes	No
Total cost	402643.0	439986.0
Total CO_2	309387.36	328291.69
Total time	63	66
Computational time (s)	2967	2952

emissions decreased from 335994.8 to 309387.36, and the time decreased from 72 to 63 days. However, increasing the tabu list size to 45 worsened the solution quality, as the cost increased from 402643.0 to 448711.0, the CO_2 emissions went from 309387.36 to 321039.72, and the time increased from 63 days to 67 days. The tabu list size parameter does not have a big influence on the computational time, it is only a matter of seconds.

Aspiration Criteria The aspiration criterion has the power to overrule a move's tabu status. In some circumstances, it may invalidate the tabu property and maintain a suitable balance between diversification and intensification to avoid certain missing solutions during the search. In the following, we fix the tabu list size to 30, the evaluation number to 2000 and we compare the results with and without aspiration criteria.

From Table 6, the researchers conclude that the aspiration criteria improved the solution quality. The aspiration criteria allowed to have better solutions as the cost of the solution improved from 439986.0 to 402643.0, the CO_2 emissions decreased from 328291.69 to 309387.36, and the time decreased from 66 days to 63 days. However, the computational time was not much influenced by the use of aspiration criteria.

The final parameters' configuration is then: a list size of 30, an enumeration number of 2000 and we fix the aspiration criteria to yes.

5.2.2 The Genetic Algorithm Parameters Exploration

As described previously, the researchers follow the same process to determine the best GA' parameters combination. The used parameter values are defined in Table 7.

Fitness Evaluation Number Similarly we test the function evaluation number and we fix the other parameter values, respectively, to 110 for the population size, 0.8 for the crossover probability and 0.2 for the mutation probability.

From Table 8, the researchers can observe that the bigger the number of evaluations is, the better the solution quality is. The cost varied from 412857.0 to 379873.0, the CO_2 went from 332958.93 to 299820.65, and the time decreased from 64 days to 58 days. For the computational time, the value increased from 3412 (s) for 1500 evaluations to 4023 (s) for 2000 evaluations number.

Population Size The ability to find the optimal solution in the search space is significantly influenced by the population size. According to numerous researches, a huge population increases the chance of discovering the optimum solution [28]. In the following experiment, the researchers fix the function evaluation to 2000, the crossover probability to 0.8, the mutation probability to 0.2 and results with three different values for the population size, equal to 90, 100, and 110, were compared.

Table 9 presents the results of varying the GA population size parameter. We can observe that the larger the population is, the better the results are. From size 90 to size 110, the cost decreased from 412857 to 328869, the CO_2 emissions went from 332958.93 to 257934.42, and the time decreased from 61 days to 55 days. However,

Table 7 GA parameters values

Parameters	Values
Evaluation number	1500
	2000
Population size	90
	100
	110
Crossover probability	0.7
	0.8
	0.9
Mutation probability	0.1
	0.2
	0.3

Table 8 GA: Comparison of the objective functions according to the function evaluation number

GA	1500	2000
Total cost	412857.0	379873.0
Total CO_2	332958.93	299820.65
Total time	64	58
Computational time (s)	3412	4023

Table 9 GA: Comparison of the objective functions according to the population size

GA	90	100	110
Total cost	412857.0	379873.0	328869.0
Total CO_2	332958.93	299820.65	257934.42
Total time	61	58	55
Computational time (s)	3801	4023	4592

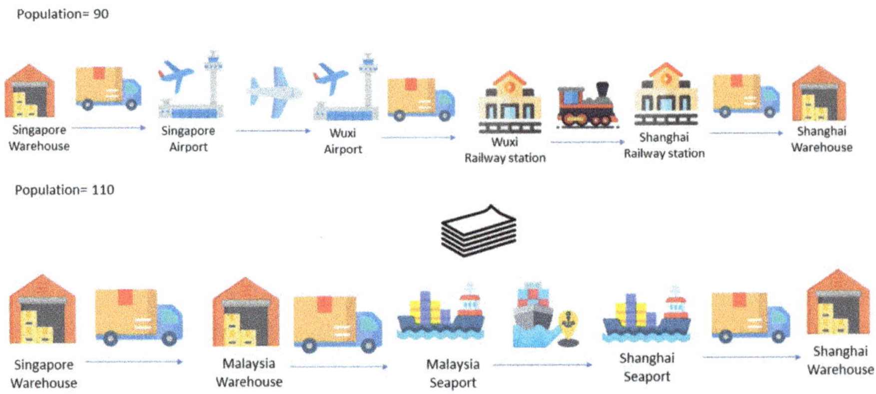

Fig. 6 Paper plates shipment from Malaysia to Singapore

the computational time increased from 3801 (s) to 4592 (s) because there are more chromosomes to operate variations on.

Figure 6 shows the paths of the paper plates shipment from Singapore to Shanghai, first with a population size equal to 90 and then with a population size equal to 110.

The population's size indicates how many chromosomes a population has (in one generation). Only a small portion of the search space is explored, if there are insufficient chromosomes for GA to perform crossover. On the other hand, GA is slowed down by an excessive number of chromosomes. Based on the findings from the literature review, it has been determined that increasing the population size beyond a certain threshold does not contribute to the acceleration of problem-solving, as the primary factors influencing this process are the encoding method and the specific problem at hand [28].

Crossover Probability The crossover and mutation probability parameters are important factors that can influence the GA results. To evaluate the influence of crossover probability PC and mutation probability Pm, the researchers consider three different couples of values presenting their combination which are (Pc = 0.7, Pm = 0.3), (Pc = 0.8, Pm = 0.2), and (Pc = 0.9, Pm = 0.1).

Table 10 GA: Comparison of the objective functions according to the crossover, population rates

GA	(0.7,0.3)	(0.8,0.2)	(0.9,0.1)
Total cost	220607.0	204026.0	204115.0
Total CO_2	38708.30	38164.4	48315.93
Total time	34	17	29

Table 11 Final configuration

Parameters	values
Genetic algorithm	
Number of population	110
Crossover probability	0.8
Mutation probability	0.2
Tabu search	
Tabu list size	30
Aspiration criteria	Yes

Table 10[3] presents the results corresponding to different combinations of crossover and mutation probabilities. It shows that increasing the crossover probability does not enhance always the solution quality. For instance, we can notice that the better combination corresponds to 0.8 for crossover and 0.2 for mutation. This combination gives the minimal cost, CO_2 emission, and time values.

To conclude, in the GA we consider the following final configuration: a population size of 110, an enumeration number of 2000 and the combination of the crossover probability PC and mutation probability Pm (0.8, 0.2).

The final configurations for TS and GA are presented in Table 11. The enumeration number is set to 2000 for the two approaches.

5.3 *Comparative Results*

To compare the two algorithms, the researchers display in Table 12 the total cost, total CO_2 emissions, total time, and computational time for each of them. These values represent the evaluation of the entire journeys for the eight freights. The researchers can conclude that GA has better performance than TS, because it can operate on more solutions and has a better balance between intensification and diversification thanks to its recombination operators. However, these operators can be costly time-wise, that is why GA has a computational time of 4592 (s) while TS has a computational time of 2967 (s). These results are confirmed by Tables 13 and 14 that present the detailed paths, respectively, by Tabu Search and Genetic Algorithm. They show the differences between these approaches for the four freights honey, paper plates, Cigarettes, and Durian, in terms of modes, path duration, cost,

[3] The obtained values correspond to the total of values obtained for the freights Honey, paper plates, Cigarettes, and Durian.

Table 12 Algorithms results

	TS	GA
Total cost	402643.0	328869.0
Total CO_2	309387.36	257934.42
Total time (days)	63	55
Total objective function	712093.36	586858.42
Computational time (s)	2967	4592

Table 13 Tabu Search comparative results for each freight

Freight	Cost	CO_2	Departure time	Deadline	Arrival time	Days	Modes
Honey	64563.0	15394.8	2018-02-01	25-Feb-18	2018-02-7	6	Truck-Truck-Sea-Truck
Paper plates	42394.0	11896.3	2018-02-03	23-Feb-18	2018-02-10	7	Truck-Air-Truck-Truck
Cigarettes	68791.0	8475.71	2018-02-05	22-Feb-18	2018-02-11	6	Truck-Air-Truck
Durian	41,136	10394.8	2018-02-07	24-Feb-18	2018-02-10	3	Truck-Rail-Truck

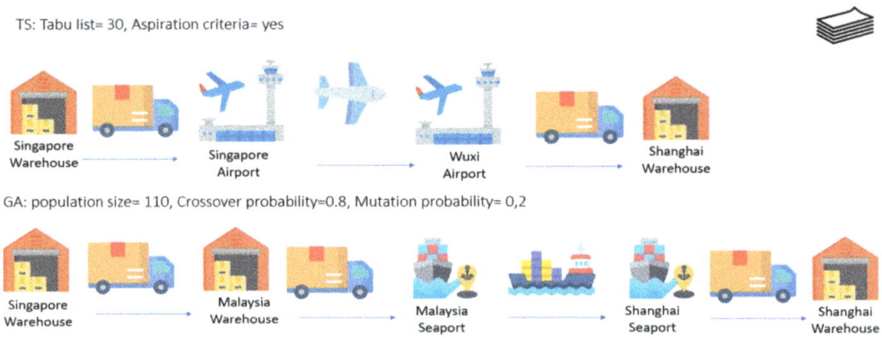

Fig. 7 Paper plates shipment path with TS, GA from Singapore to Shanghai

and CO_2 emission. Moreover, Fig. 7 shows the improvement of the paths of the paper plates shipment using the TS and GA.

6 Conclusion

The researchers propose a new model, closer to the reality, that takes into consideration four different modes (air, road, rail, and sea) simultaneously. The multimodal, multi-objective, and multi-commodity model takes into account three non-conflicting aggregated objectives that are cost, CO_2 emissions, and time. The researchers adopted two algorithms TS and GA to solve the MFT route planning

Table 14 Genetic algorithms comparative results for each freight

Freight	Cost	CO2	Departure time	Deadline	Arrival time	Days	Modes
Honey	55964.0	13563.2	2018-02-01	25-Feb-18	2018-02-5	4	Truck-Truck-Sea-Truck
Paper plates	41639.0	9968.3	2018-02-03	23-Feb-18	2018-02-07	4	Truck-Truck-Sea-Truck
Cigarettes	66968.0	6864.7	2018-02-05	22-Feb-18	2018-02-13	8	Truck-Sea-Truck-Truck
Durian	39455.0	7766.2	2018-02-07	24-Feb-18	2018-02-08	1	Truck

problem. The experimental results showed that the GA performed better than TS which explains the driving force of a population-based metaheuristic. The GA algorithm solution reduced the overall cost, CO_2, and time, respectively, by, 73774, 51472.94, and 8. The resulting solution provides us with the appropriate paths to ship the goods from their source to their destination, respecting certain constraints like time and feasibility. These routes are recommended to the carriers in order to know which mode to use. We obtain than an economic, ecological, and green freight transportation that guarantees the client satisfaction in terms of delivery time. However, these results are obtained under a number of limiting hypotheses.

As future work, we plan first to compare our results with those of other approaches such as hybrid metaheuristics to benefit from their advantages and also with those of other approaches proposed in the literature. Second, we plan to integrate uncertainty in our model to take into consideration other factors such as congestion, natural disasters, or accidents. These factors can lead to disruptions and delays that need to be taken into account. Third, we aim to build and optimize fully synchronized systems as the different modes, apart trucks, have their own schedules. Finally, as the type of goods has an influence on the freight routing, the researchers will deal with types of freight that need more care which leads to more constraints. For example, carrying food can be very challenging for the carriers. They have to consider special containers and special modes. Moreover, the time constraint will become more restricted.

References

1. C. Archetti, L. Peirano, M.G. Speranza, Optimization in multimodal freight transportation problems: A survey. Eur. J. Oper. Res. **299**(1), 1–20 (2022)
2. M. SteadieSeifi, N.P. Dellaert, T.W. Nuijten, T. Van Woensel, R. Raoufi, Multimodal freight transportation planning: a literature review. Eur. J. Oper. Res. **233**(2), 1–15 (2014).
3. N. Zufferey, M. Verma, Tabu Search for an Intermodal Transportation Problem, in *Selected Logistics Problems and Solutions* (2011), pp. 7–20

4. D.C. Pereira, A.G. del Valle, R.R. Prado, N.R. Monteil, D. del Rio Vilas, Hybrid algorithm for the optimization of multimodal freight transport services: Maritime Application, in *2013 Winter Simulations Conference (WSC)* (IEEE, New York, 2013), pp. 3406–3417
5. K. Lei, X. Zhu, J. Hou, W. Huang, Decision of multimodal transportation scheme based on swarm intelligence, in *Mathematical Problems in Engineering* (2014)
6. M. Hrušovský, E. Demir, W. Jammernegg, T. Van Woensel, Hybrid simulation and optimization approach for green intermodal transportation problem with travel time uncertainty. Flex. Serv. Manuf. J. **30**, 486–516 (2018)
7. Y. Sun, M. Hrušovský, C. Zhang, M. Lang, A time-dependent fuzzy programming approach for the green multimodal routing problem with rail service capacity uncertainty and road traffic congestion. Complexity **2018** (2018). https://doi.org/10.1155/2018/8645793
8. Y. Zhao, Q. Xue, Z. Cao, X. Zhang, A two-stage chance constrained approach with application to stochastic intermodal service network design problems. J Adv Transp **2018**, 1–18 (2018)
9. H. Heggen, Y. Molenbruch, A. Caris, K. Braekers, Intermodal container routing: integrating long-haul routing and local drayage decisions. Sustainability **11**(6), 1634 (2019)
10. Y. Sun, X. Li, X. Liang, C. Zhang, A bi-objective fuzzy credibilistic chance-constrained programming approach for the hazardous materials road-rail multimodal routing problem under uncertainty and sustainability. Sustainability **11**(9), 2577 (2019)
11. X. Zhang, F. Jin, X. Yuan, H. Zhang, Low-carbon multimodal transportation path optimization under dual uncertainty of demand and time. Sustainability **13**, 8180 (2021). https://doi.org/10.3390/su13158180
12. D. Ge, Optimal path selection of multimodal transport based on Ant Colony Algorithm, in *Journal of Physics: Conference Series*, vol. 2083(3) (IOP Publishing, New York, 2021), p. 032011
13. M. Oudani, A simulated annealing algorithm for intermodal transportation on incomplete networks. Appl. Sci. **11**(10), 4467 (2021)
14. Y. Sun, N. Yu, B. Huang, Green road–rail intermodal routing problem with improved pickup and delivery services integrating truck departure time planning under uncertainty: an interactive fuzzy programming approach. Complex Intell. Syst. **8**, 1459–1486 (2022). https://doi.org/10.1007/s40747-021-00598-1
15. Y. Lu, F. Chen, P. Zhang, Multiobjective optimization of multimodal transportation route problem under uncertainty. Eng. Lett. **30**(4) (2022). https://doi.org/10.1016/j.jrtpm.2023.100381
16. M. Li, X. Sun, Path optimization of low-carbon container multimodal transport under uncertain conditions. Sustainability **14**, 14098 (2022). https://doi.org/10.3390/su142114098
17. S. Okyere, J. Yang, C.A. Adams, Optimizing the sustainable multimodal freight transport and logistics system based on the genetic algorithm. Sustainability **14**(18), 11577 (2022)
18. L. Yang, C. Zhang, X. Wu, Multi-objective path optimization of highway-railway multimodal transport considering carbon emissions. Appl. Sci. **13**(8), 4731 (2023). https://doi.org/10.3390/app13084731
19. H. Huang, D. Bucher, J. Kissling, R. Weibel, M. Raubal, Multimodal route planning with public transport and carpooling. IEEE Trans. Intell. Transp. Syst. **20**(9), 3513–3525 (2018).
20. D. Chen, Y. Zhang, L. Gao, R.G. Thompson, Optimizing multimodal transportation routes considering container use. Sustainability **11**(19), 5320 (2019)
21. S. Yang, Z. Song, Multi-objective optimization model of multi-modal transport based on regional sustainability indicators, in *MATEC Web of Conferences*, vol. 325 (EDP Sciences, New York, 2020), p. 03001
22. J.H. Holland, *Adaptation in Natural and Artificial Systems: An Introductory Analysis with Applications to Biology, Control, and Artificial Intelligence* (MIT Press, New York, 1992)
23. F. Glover, M. Laguna, R. Marti, Principles of tabu search. Approximation Algorithms and Metaheuristics **23**, 1–12 (2007)
24. K. Sastry, E.D. Goldberg, Modeling tournament selection with replacement using apparent added noise, in *Intelligent Engineering Systems Through Artificial Neural Networks*, vol. 11 (2002), pp. 129–134

25. D.A. Coley, *An Introduction to Genetic Algorithms for Scientists and Engineers* (World Scientific Publishing Company, Singapore, 1999)
26. A. Hassanat, E. Alkafaween, On enhancing genetic algorithms using new crossovers. Int. J. Comput. Appl. Technol. **55**(3), 202–212 (2017)
27. S.S. Alves, S.A. Oliveira, A.R.R. Neto, A novel educational timetabling solution through recursive genetic algorithms, in *IEEE 2015 Latin America Congress on Computational Intelligence (LA-CCI)* (2015), pp. 1–6
28. B.R. Rajakumar, A. George, APOGA: An adaptive population pool size based genetic algorithm. AASRI Procedia **4**, 288–296 (2013)

Solving Hierarchical Production–Distribution Problem Based on MDVRP Under Flexibility Depot Resources in Supply Chain Management

Abir Chaabani and Lamjed Ben Said

1 Introduction

Bi-level optimization, as the name reflects, deals with decentralized problems that feature interactive decision entities distributed throughout two levels of hierarchy: (1) a leader, called the upper-level problem, and (2) a follower, called the lower-level problem. In such situation, the lower-level problem appears as a constraint, such that only an optimal solution to the lower-level problem is a possible feasible candidate to the upper-level one. Each level or decision-maker tries to optimize its own objective function without considering the objective of the other part, but the decision of each part affects the objective value of the other one as well as the decision space, which makes bi-level optimization problems difficult to handle [1]. The particular structure of these problems facilitates the formulation of several practical situations that involve a nested decision-making process. For this reason, a large broad of real-life applications can be cast within this hierarchical decision-making structure such as: supply chain [2], transportation [3], program debugging [4], chemical engineering [5], cloud computing [6], assignment application [7], etc.

In this chapter, we are motivated by supply chain systems that require finding optimal decision at various stages of the whole process. Such decisions may be related to the location of manufacturing plants or distribution centers, the acquisition of raw materials, the production process, the inventory control, the suitable delivery of commodities, etc. Particularly, one of the main topics discussed in supply chain is the production–distribution problem that involves multiple decision-makers often connected by means of hierarchical links. These systems are related in nature; however, in most practical situations and research works, each decision entity

A. Chaabani (✉) · L. Ben Said
SMARTLab, ISG, Tunis, Tunisia
e-mail: abir.chaabani@isg.rnu.tn

I. Alharbi et al. (eds.), *Advances in Computational Logistics and Supply Chain Analytics*, Unsupervised and Semi-Supervised Learning,
https://doi.org/10.1007/978-3-031-50036-7_6

concentrates on optimizing its process independently to other related decisions. Both, in general, do not cooperate because of different optimization strategies. Such characteristics render these problems unsuitable for modeling by standard mathematical programming. For this reason, researchers have proposed new formulations for the production–distribution planning based on two different decision-makers controlling both production and distribution process with a hierarchical relationship between them [8–10]. This might be systems in which a principal firm, at the upper level of the decision process, controls the distribution centers and seeks to minimize the transportation costs. At the lower level, a manufacturing plant, on receiving the order of the upper company, seeks to minimize its operating costs. In this way, a distribution company owns a homogeneous fleet of vehicles based at different depots and has to serve a set of retailers geographically dispersed with known demands at the upper level. This company decides on the depot that serves each retailer and designs the routes that serve them. In addition, it is responsible for providing sufficient quantities of products to each depot in order to satisfy the demands of the customers assigned to that depot. Thus, to be able to reduce the overall cost of products, the distribution company should not only minimize the cost of serving items from depot to retailers, but it should also take into account the cost of acquiring the products and unloading them at the depots. In this way, a production company at the lower level controls a set of manufacturing plants, receives order from the distribution company, and allocates the production to its plants. The aim is to minimize the operation costs. In this model, the upper-level problem can be seen as a multi-depot vehicle-routing problem (MDVRP), and the whole problem is formulated as a Bi-MDVRP, which is an extension of the well-known vehicle -routing problem (VRP).

In this chapter, we adopt this description and we added constraints related to the flexibility of choice of the stop depot. In fact, the baseline variant of the problem considers only a vehicle capacity constraint and a fixed depot position that each vehicle should start and end the route at this position in the upper-level problem. Thus, to improve the cost of generated solution and to improve the satisfiability of customers, we added this constraint to the model that we called Bi-MDVRPFD (a Bi-level Multi-Depot VRP under Flexibility Depots). In order to solve the bi-level program proposed to formulate the hierarchical PD problem, we customize and adapt a co-evolutionary decomposition-based approach called CODBA (Co-evolutionary Decomposition-Based Algorithm) [11]. Our motivation behind the use of this approach is justified by the fact that CODBA is a recently proposed bi-level solution approach that respects the nested bi-level structure (it does not modify the structure of the problem), and it applies a co-evolutionary decomposition scheme at the lower-level problem to minimize the computational overhead of the follower evolutionary algorithm. CODBA exploits three main mechanisms that are decomposition, multi-threading, and co-evolution within the lower level . The idea is to use a decomposition-based method as a surrogate to the lower-level complexity. In this way, the proposed approach decomposes the lower-level population into a set of well-distributed sub-populations using a new proposed method for discrete decision space decomposition called discrete space decomposition method (DSDM).

The generated sub-populations co-evolve simultaneously in a parallel manner using multiple threads and exchange information via recombination with the best lower-level individuals. To this end, the main contributions of this work are summarized as follows:

1. Proposing a new bi-level formulation for production–distribution planning system based on flexibility depot position constraints
2. Developing the CODBA algorithm to our problem to obtain optimal cost solution
3. Reporting experimental results with respect to the competitor algorithm

The rest of this chapter is organized as follows. Section 2 presents the bi-level optimization background. Section 3 describes the hierarchical PD planning system with flexibility depot resources. Section 4 goes on to develop the method proposed to solve the problem. Section 5 gives the experimental results in this chapter. Finally, Sect. 5 concludes the work and provides future perspectives.

2 Bi-level Optimization: Basic Concepts

BLOPs appeared as an optimization task containing a nested inner optimization task as a constraint of an outer optimization problem. The latter is commonly referred as the leader's (upper level) optimization problem, and the inner optimization task is known as the follower's (or lower level) optimization problem. The two involved DMs have different priorities on decision in the sense that the realized outcome of any decision taken by the upper to optimize its goal is affected by the response of lower-level entity. It is assumed that the upper-level DM, who has higher priority in making its own decision, first specifies a strategy. Then, the lower-level DM reacts with full knowledge of the leader's decision. Accordingly, the leader's solution is implicitly affected by the follower's reaction. In this hierarchical problem, decision variables are split into two groups that are controlled by two decision-makers. The latter have their own objectives and constraints. Indeed, there are coupling constraints that connect the decision variables of the leader and the follower. In the following, we present the general problem formulation.

Definition 1 (Bi-level Optimization Problem Formulation) BLOP is formulated as

$$\underset{x_u \in X_U, x_l \in X_L}{Min} \quad F(x_u, x_l) \quad . \tag{1}$$
$$s.t. \begin{cases} x_l \in ArgMin\{f(x_u, x_l),\} \, g_i(x_u, x_l) \leq 0, \\ G_j(x_u, x_l) \leq 0, \; i = 1, ..., I \; and \; j = 1, ..., J \end{cases}$$

In this formulation, two types of decision variables are handled: (1) the upper-level variables $x_u \in X_u$ and (2) the lower-level variables $x_l \in X_L$. For the follower problem, the optimization task is performed with respect to the variables x_l, and the variables x_u act as parameters. For each x_u corresponds a different follower problem, whose optimal solution needs to be determined. All variables (x_u, x_l) are considered in the leader problem, and the optimization is expected to be performed with respect to both sets of variables. The function $G_j : X_U \times X_L \rightarrow \mathbb{R}$ denotes the upper-level constraint, and $g_i : X_U \times X_L \rightarrow \mathbb{R}$ represents the lower-level constraint, respectively. The difficulty in bi-level optimization arises from the fact that only lower-level optimal solutions could be considered as feasible members to the upper level after satisfying the upper constraints. For instance, a member $x^1 = (x_u^1, x_l^1)$ is considered feasible at the upper-level problem only if x^1 satisfies the upper-level constraints, and x_l^1 is an optimal solution regarding to the lower-level problem corresponding to x_u^1.

The decision variables x_u and x_l could be continuous, discrete, or even mixed. In this chapter, we are interested in problems where variables of both levels are discrete. Moreover, the presented definition of bi-level optimization corresponds to the optimistic case. In fact, when the lower- level problem has multiple optimal solutions, there is lack of clarity as to which lower-level optimal solution should be used at the upper level. It is common to take either of the two positions, i.e., optimistic or pessimistic, to sort out this ambiguity. In the optimistic case, some form of cooperation is assumed between the upper and lower levels where the latter is expected to choose the optimal solution that leads to the best objective function value for the upper-level problem. However, in a pessimistic position, the upper level optimizes for the worst case where the lower level may choose a solution from the optimal set that leads to the worst objective function value for the upper level.

3 Hierarchical Production–Distribution Problem Under Flexibility Depot Resources

3.1 PD-Related Works

The consideration of an integrated production–distribution system in supply chain management was discussed in the literature by a number of authors [12, 13]. These studies assume a principal firm having multiple plants and distribution centers that control the integrated PD process or multiple firms that are assumed to act simultaneously since they collaborate to achieve common goals.

The authors in [14] explore the integration of production and distribution planning in a supply chain context, considering the flexibility of depot operations. The study proposes models and optimization algorithms to optimize production and distribution decisions while accounting for various depot-related flexibilities.

The authors in [15] focus on optimizing the design of production–distribution networks by incorporating the flexibility of depot capacities. The study develops mathematical models and solution approaches to determine the optimal network configuration that can adapt to changing production and distribution requirements.

The authors in [16] present a multi-objective optimization model for production–distribution planning, taking into account the flexibility of depot operations. The study aims to find the trade-off between conflicting objectives, such as cost minimization, resource utilization, and service-level improvement, while considering different depot-related flexibilities.

Liu et al. [17] focus on developing optimization models for integrated production–distribution planning, considering the flexibility of depot resources. The study proposes mathematical formulations and solution methods to optimize production and distribution decisions, considering depot-related flexibilities such as variable capacities and adaptable workforce. All the presented works treated the distribution–production problems in supply chain management under the flexibility resources using a mono-level scheme. However, in reality these systems involve decision at two different levels in a hierarchical way. For this reason, it is more interesting to model the problem using the bi-level framework that has been proposed in the literature as an appropriate model for hierarchical decision processes.

Cao and Chen [18] have proposed a mixed integer model describing in the upper level a manufacturing entity that decides on the opening and closing of existing plants. The upper-level entity makes his decision based on the cost of opening new plants and their production capacities. At the lower level, the decisions are to minimize the plants' operation costs. This formulation was solved using the classical resolution methods by transforming the bi-level model in a single-level decision using the Karush–Kuhn–Tucker conditions of the lower-level problem. Such transformation cannot be applied with mode due to the presence of many lower-level Lagrange multipliers.

Marinakis and Marinaki [19] formulate a location-routing problem as a bi-level problem. The upper level corresponds to the higher strategic level that decides on the location of the facilities and the lower level decides the routes that serve the customers. The authors adopted a genetic algorithm with local search procedure to solve the model. Calvete and Gale [20] suggested a hierarchical version of the PD problem using a distribution entity at the upper level to control a number of depots that acquires goods from a company owning a number of plants and then serves them to their retailers. Thus the upper-level decision entity is the distribution company, and the lower level is handled by the manufacturing company. The bi-level model is solved using the ant colony optimization meta-heuristic.

3.2 Mathematical Formulation

In this section, we describe the general characteristics of our formulation, and we present the detailed mathematical model. Based on Calvete et al. [20] problem's description, it assumes to be a planning process involving two different decision-makers (DMs). In the upper level, a distribution entity controls the routing problem and is formulated based on the MDVRP model. In this way, the distribution

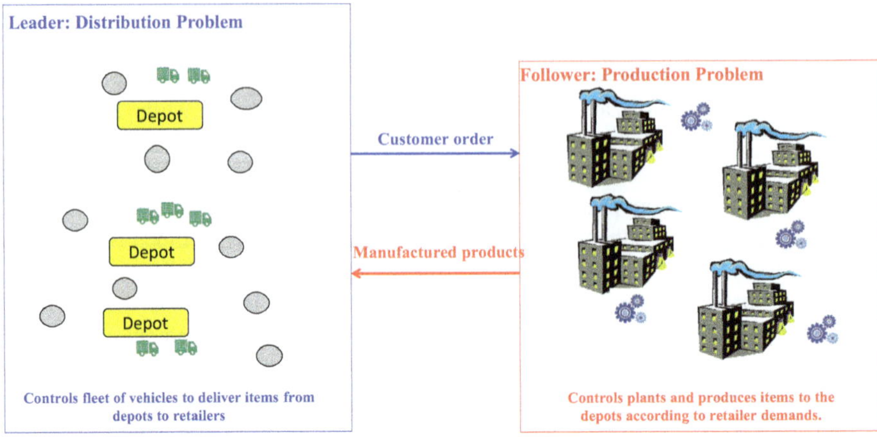

Fig. 1 A bi-level modeling of the multi-objective production–distribution planning system in supply chain management

company has to serve a set of retailers geographically dispersed with known demands. This decision level designs the routes for serving their customers with the aim to minimize the distribution items' cost. In order to satisfy the demands of customers, a lower decision level allocates the production to its plants with the aim to minimize the production cost. To this end, the overall cost of served item depends on both routing and manufacturing costs (cf. Fig. 1). In the baseline problem formulation, each vehicle must start and end at the same depot. In this manner, the delivery person should return to his starting point even if he is close to another delivery depot. In this contribution, we study a new variant called collaborative transportation, in which the vehicles may end at the depot nearest to the last customer to improve logistic performance. Consequently, the upper-level problem is termed MDVRPFD and the whole problem is called Bi-MDVRPFD. We present in the following the mathematical formulation of the bi-level PD planning problem. In this model, a number of assumptions are taken into consideration:

1. Each vehicle starts from a depot and ends at the depot closest to the last customer it serves.
2. Each customer is served exactly once by one vehicle.
3. The customer demand along the route does not exceed the vehicle capacity.
4. All of the customers with known demands are assigned to vehicles.
5. The number of vehicles returning to a depot cannot exceed the number of parking spaces at a depot d.

$$\underset{x_{ij}^{s},y_{kl},}{Minimize}: \sum_{s\in S}\sum_{(i,j)\in E} c_{i,j}^{a}x^{s}_{i,j} + \sum_{l\in L}\sum_{k\in K} c_{k,l}^{b}y_{k,l} \quad . \tag{2}$$

$$s.t.\begin{cases}
\sum_{s\in S}\sum_{j\in R} x_{ij}^{s} = \sum_{s\in S}\sum_{i\in E} x_{ij}^{s} = 1 \;\; \forall i \in R & 1(a)\\
\sum_{d\in D}\sum_{j\in R} x_{dj}^{s} = \sum_{d\in D}\sum_{i\in E} x_{id}^{s} \leq 1 \;\; \forall s \in S & 1(b)\\
\sum_{i\in R}\sum_{j\in R} x_{ij}^{s} \leq |R| - 1, \;\; \forall R \subseteq E, \; s \in S & 1(c)\\
\sum_{i\in D}\sum_{j\in D} x_{ij}^{s} = 0, \;\; \forall s \in S & 1(d)\\
\sum_{s\in S}\sum_{i\in R} x_{id}^{s} \leq |p_{d}|, \;\; \forall d \in D & 1(e)\\
\sum_{s\in S}\sum_{i\in R} x_{id}^{s} \geq 1, \;\; \forall d \in D & 1(f)\\
\sum_{i\in E}\sum_{j\in R} q_{j}x_{ij}^{s} \leq Q_{s} \;\; \forall s \in S & 1(g)\\
x_{ij}^{s} \in \{0, 1\}, \; (i, j) \in E, \; s \in S & 1(h)\\
where, \; for \; given \left\{x_{ij}^{s}\right\}, \; \left\{y_{k,l}\right\} solves: & \\
\quad \underset{y_{k,l}}{min} \sum_{k\in K} c_{k,l}^{c}y_{k,l} & 1(i)\\
\quad s.t. \sum_{l\in L} y_{k,l} \leq A_{k}, \; k \in K & 1(j)\\
\quad \sum_{k\in K} y_{k,l} \geq \sum_{s\in S}\sum_{r\in R} q_{r}, l \in L & 1(k)\\
\quad y_{k,l} \geq 0 & 1(l)
\end{cases}$$

In the following, we present the used notations of the problem formulation.

K: set of plants
R: set of retailers
S: set of vehicles
D: set of depots
E: set of edges connecting depot to retailers
qr: the demand of retailer r
A_k: the production availability of plant k
p_d: number of parking places at depot d
Q_s: total load
A_k: plant production

where: c_{ij}^{a}, $c_{k,l}^{b}$ and $c_{k,l}^{c}$ are the cost of transporting goods from i to j, $(i; j) \in E$, the cost of acquiring and unloading a product manufactured in plant k into depot l and the cost of manufacturing a product at plant k for depot l, respectively. In this model, the upper constraints are defined as follows:

- Constraint 1(a) ensures that each customer must be visited exactly once by exactly one vehicle.

- Constraint 1(b) represents that each vehicle starts from a depot and ends at a depot.
- Constraint 1(c) is the subtour elimination constraint.
- Constraint 1(d) imposes that the vehicle cannot travel directly from depot i to depot j.
- Constraint 1(e) imposes that the number of vehicles returning to each depot is not more than the number of available parking places $|p_d|$.
- Constraint 1(f) imposes that the number of vehicles at each depot is at least equal to 1.
- Constraint 1(g) defines that the total demand of all retailers on one particular route must not exceed the capacity of the vehicle assigned to this route.
- Constraint 1(h) imposes variable binary requirements.

In the lower level, we consider the following constraints:

- Constraint 1(j) guarantees that production availability of plant k should not be exceeded.
- Constraint 1(k) ensures that demands of retailers should be satisfied by associating a part of plant production to depots.
- Constraint 1(l) imposes variable the positivity requirement.

4 A Co-evolutionary Decomposition-Based Algorithm for the Bi-MDVRPFD

In this section, we describe the Co-evolutionary Decomposition-Based Algorithm (CODBA) in solving the bi-level MDVRPFD. The latter is a recently proposed bi-level method designed to solve efficiently bi-level combinatorial problem [7]. CODBA uses three main mechanisms at the lower level that are: (1) decomposition, (2) multi-threading, and (3) co-evolution. The first mechanism consists in decomposing the population into several well-distributed sub-populations over the search space in order to improve the diversity rate of the algorithm. Each sub-population is responsible to explore a specific region of the search space. Then, the best global solution is chosen based on the sub-populations' optima. These sub-populations are evolved in parallel using several threads such that each sub-population is assigned to a distinct thread. In this way, each cluster (thread) applies a genetic algorithm (GA) with the aim to find the clusters' optimal solutions. When the evolution process of the different sub-populations terminates, we store their best solutions into an archive. The obtained best individuals are subsequently exchanged between the different parts via the crossover operation. In fact, making variation with the archive solutions allows information exchange between the sub-populations. The step-by-step upper and lower procedures of the our developed algorithm version on the Bi-MDVRPFD are described in the following:

Upper Level: Distribution Problem

[$\mathbf{D_1}$, 1, 2, 3, -4, 9 $\mathbf{D_2}$, 5, 6, $\mathbf{D_3}$, 7, 8, $\mathbf{D_3}$]

Lower Level: Production Problem

	P_1	P_2	P_3
D_1	90	0	0
D_2	40	0	0
D_3	20	10	35

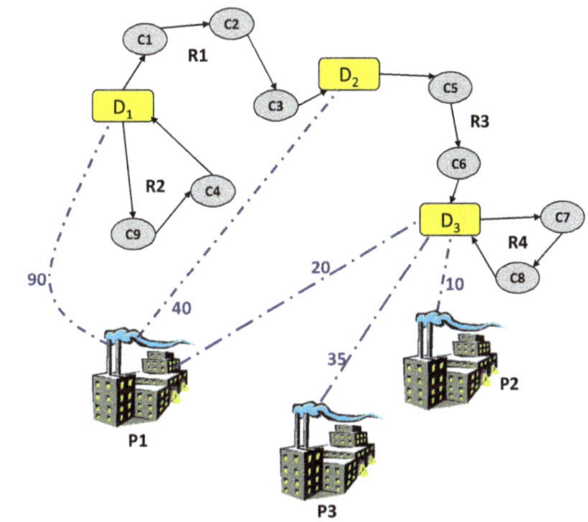

Fig. 2 Solution encoding for an example of Bi-MDVRPFD with 9 retailers, 3 depots, and 3 plants

4.1 Upper-Level Procedure

- **Step 1 (Initialization scheme)**: We generate an initial parent population of N solutions randomly. Then, the lower-level optimization procedure is executed to identify the optimal individuals according to such generated upper solution. In fact, the upper-level fitness is assigned based on both upper-level function value and constraints (described by Eq. (2)). In this chapter, our encoding upper solution is represented using a vector. Each route is a sequence of indices from 1 to n (n is the number of customers) delimited by the starting and/or ending depot. Figure 2 illustrates an example of solution encoding for an instance of Bi-MDVRPFD. The upper solution for 9 retailers and 3 depots (corresponding to the distribution problem) defines 4 routes. The vehicles at depot D_1 start serving customer 1, 2, and then customer 3. After that, a new route from the same depot is dedicated to serve customer 4 and then customer 9. In this way, the negative sign presents the starting of a new route from the same depot (starting depot). After completing its tour, the used vehicle returns to the depot 2, which is the nearest depot to customer 3. Similarly, the vehicle started from depot 2 returns to D_3 after serving customers 5 and 6. The depot 3 represents here the nearest depot to the customer 6. Lastly, the vehicle at depot 3 serves customers 7 and then 8 and returns to the depot 3 after satisfying the customer demands.
  We note here that regarding to the upper solution feasibility, vehicle chooses to return to a depot D_i only when constraints related to the number of used vehicles and parking places at this position are not violated.
- **Step 2 (Upper-level parent selection)**: We choose $(N/2)$ population members from the parent population using tournament selection operator to participate to

the mating pool. This set contains N/2 individuals that serve for reproduction, passing their genetic information to the next generation.

- **Step 3 (Variation at the upper level)** Perform crossover and mutation operators in order to create the offspring population. We used a combination of two variation operators:

  1. RBX [21] that is a well-used operator with the routing problem. The latter copies routes from a parent solution and then completes the offspring with routes from the other parent by removing visited retailers.
  2. UMutation [22] replaces the value of a chosen customer with a uniform random value selected from the decision range interval.

- **Step 4 (Lower-level optimization)** Solve the lower-level optimization problem for each upper offspring member using the decomposition-based co-evolutionary parallel scheme.
- **Step 5 (Offspring evaluation)** Combine both the upper-level parents and the upper-level children into an R_t population and evaluate them based on the upper-level objective function and the constraints.
- **Step 6 (Environmental selection)** Fill the new upper-level population with the N best solutions of R_t. If the stopping criterion is met, then return the best upper-level solution; otherwise, return to Step 2.

4.2 Lower-Level Procedure

- **Step 1 (Lower-level decomposition)** For each upper-level solution, we generate a number of well-distributed structured points P on the whole discrete decision space using a decomposition method called Discrete Space Decomposition Method (DSDM) [7]. The latter generates a number of sub-populations over the lower decision space. Each sub-population member is evaluated using the lower-level objective function and constraints Eq. (2). We note here that we choose for lower solution encoding a matrix M of real numbers with $K; L$ dimensions, such that $M[k; l]$ defines the quantity of articles produced by the plant k for depot l according to a fixed demand.
- **Step 2 (Lower-level parent selection)**: We choose $SPS/2$ population members from each lower-level parent sub-population using tournament selection where SPS is the lower sub-population size.
- **Step 3 (Lower-level parent variation)**: Perform the single-point crossover and the bit uniform mutation for crossover and mutation operators, respectively.
- **Step 4 (Lower-level offspring evaluation)**: Combine each parent sub-population with its corresponding offspring population and evaluate them using the lower-level objective function and constraints [presented in Eq. (2)].
- **Step 5 (Lower-level environmental selection)**: Fill each new lower-level sub-population using a replacement strategy. In fact, each new lower-level sub-population is formed with the SPS best solutions of the combined one. If the

stopping criterion is met, then store the best found lower-level solution in the archive; otherwise, return to step 2.
- **Step 6 (Co-evolution)**: The archive contains the best found solutions during the evolution. Thus, to have a global view of the whole search space and to be able to determine the global optimal solution, a co-evolution step is required. Each sub-population member is crossed-over with one of the best archive members of the other sub-populations. This process is repeated for $MaxGenCoEvol$ where $MaxGenCoEvol$ is the maximum allowed number of generations for co-evolution. It is important to note that all sub-populations are to be evolved simultaneously using a thread for each sub-population.

Feasibility and optimality of the resolution step are treated mainly by the CODBA scheme in which the decomposition and co-evolution mechanisms promote the quality of solution in terms of diversity and convergence. Moreover, the multi-threading mechanism improves the execution time of the algorithm that promotes feasibility.

5 Experimental Study

The motivation behind this chapter is to solve a new formulation of the production–distribution problem using a recently proposed bi-level solution method. In this section, we first present the empirical study design: the used benchmark problems, the performance indicators, and the used parameters setting. Then, we report a comparative experiment against:

- CODBA [7] a cooperative bi-level algorithm solving the baseline model
- Repair method [23] that is a nested bi-level method that uses evolutionary algorithm (GA) at both levels to handle bi-level optimization problems

We note that both used algorithms apply the same components: the stopping criterion, the variation operators, and the initializers. Moreover, CODBA and repair are coded in Java programming language, and all simulations are performed on the same machine ($Intel®Xeon®ProcessorCPUsE5 - 2620v3, 16\,GBRAM$).

5.1 Benchmark Problems

In this subsection, we describe the different benchmark problems used in our experimental study. In fact, we have adopted two suites of test problems ((1) the Bi-p suite and (2) the Bi-pr one) that have been derived from 33 published MDVRP data sets. It is important to note that the Bi-pr suite instances are collected by Cordeau et al. [24]. Besides, the bi-p suite instances are mainly extracted from real cases of Cordeau [25]. Particularly, bi-p1–7 are collected from the work of Christofides and

Eilon [26], bi-p8–11 are taken from the work of Gillet et al. [27], and bi-p12–23 are extracted from the work of Chao et al. [28].

We note that there is no benchmark instance set that is available for Bi-MDVRPFD as it is a new variant of MDVRP based on a bi-level production–distribution scheme. To this end, we construct a set of instances based on the baseline benchmarks and following the modus operandi described in Sect. 3:

- We add plants as many as there are depots randomly located on the map.
- We derived the number of parking places based on the number of used vehicles at each depot as presented in [29].

Indeed, the description properties of the instances are given in Table 1, and the other details are available via the link.[1] In Table 1, R denotes the number of customers that need to be served, C represents the capacity of the vehicles in each instance, S denotes the number of vehicles, P_d represents the number of parking spaces, D represents the number of depots, and P is the number of plants for the lower level. Regarding the maximal production parameter of the different plants, we set the maximal value in order to ensure the instance feasibility.

5.2 *Parameter Setting*

This section is devoted to detail the parameter setting for both used algorithms under comparison. To achieve more stable and accurate results, we considered the same terminal criteria represented by the same number of function evaluations (FEs) as termination criteria, which is set to 2500, for both levels. The F_{dec} decomposition factor used by CODBA in the lower is determined based on the problem dimension and is derived from the work of [11]. The other commonly used parameters setting is as follows: population size, crossover probability, and mutation probability, which are set to 100, 0.9, and 0.1, respectively.

Regarding the parameters of the bi-level production–distribution system instances, we use: an operation cost $c^c_{k;l}$ ranges from 1 to 4, a buy and unload cost $c^b_{k;l}$ ranges from 0.36 to 5.36, and a production availability chosen from the $[\frac{D}{|K|}, D]$ interval, where D is the total demand of retailers and $|K|$ represents the cardinality of the set of plants. These presented values are deduced from the work of [20]. Finally, the transportation cost $c^a_i j$ and the demand of retailers are described based on the $link^1$.

[1] https://github.com/fboliveira/MDVRP-Instances/blob/master/DESCRIPTION.md.

Table 1 Description of the used Bi-MDVRP benchmark problems

Bi-MDVRPFD instances		R	S	P_d	C	D	P
Bi-P Suite	**bi-p01-FD**	50	4	8	80	4	4
	bi-p02-FD	50	2	4	160	4	4
	bi-p03-FD	75	3	6	140	5	5
	bi-p04-FD	75	8	16	100	2	2
	bi-p05-FD	100	5	10	200	2	2
	bi-p06-FD	100	6	12	100	3	3
	bi-p07-FD	249	4	8	100	4	4
	bi-p08-FD	249	14	28	500	2	2
	bi-p09-FD	249	12	24	500	3	3
	bi-p10-FD	249	8	16	500	4	4
	bi-p11-FD	80	6	12	500	5	5
	bi-p12-FD	80	5	10	60	2	2
	bi-p13-FD	80	5	10	60	2	2
	bi-p14-FD	80	5	10	60	2	2
	bi-p15-FD	160	5	10	60	4	4
	bi-p16-FD	160	5	10	60	4	4
	bi-p17-FD	160	5	10	60	4	4
	bi-p18-FD	240	5	10	60	6	6
	bi-p19-FD	240	5	10	60	6	6
	bi-p20-FD	240	5	10	60	6	6
	bi-p21-FD	360	5	10	60	9	9
	bi-p22-FD	360	5	10	60	9	9
	bi-p23-FD	360	5	10	60	9	9
Bi-pr Suite	**bi-pr01-FD**	48	1	2	200	4	4
	bi-pr02-FD	96	2	4	195	4	4
	bi-pr03-FD	144	3	6	190	4	4
	bi-pr04-FD	192	4	8	185	4	4
	bi-pr05-FD	240	5	10	180	4	4
	bi-pr06-FD	288	6	12	175	4	4
	bi-pr07-FD	72	1	2	200	6	6
	bi-pr08-FD	144	2	4	190	6	6
	bi-pr09-FD	216	3	6	180	6	6
	bi-pr10-FD	288	4	8	170	6	6

5.3 Adopted Statistical Methodology

In order to compare between the used algorithms, we need a statistical test that provides a system for making quantitative decisions about the process. In this way, we used in a pairwise fashion the Wilcoxon rank test [30] in order to check whether the obtained results between samples are statistically different or not. We perform 31 runs for each couple (algorithm, problem instance) as recommended in [30]. We then obtain for each experiment 31 metric values for each algorithm. We use the

Wilcoxon rank sum test with the MATLAB framework in order to compute the *p-value* of the following hypothesis:

1. H_0: median (Algorithm1) = median (Algorithm2)
2. H_1: median (Algorithm 1) $\neq$ median (Algorithm 2)

for a confidence level of 95%, if the *p-value* is found to be less or equal than 0.05, we reject H_0 and we accept H_1. In this way, we can say that the distribution results of the two algorithms are different from each other and that one algorithm outperforms the other viewpoint used metric. However, if the *p-value* is found to be greater than 0.05, then we accept H_0 and we cannot say that one algorithm is better than the other nor the opposite.

5.4 Comparative Results

This section is devoted to answer the following research question:

How does CODBA based on Bi-MDVRPFD perform first compared to the nested algorithm and regarding the baseline line variant (i.e., without depot flexibility) in a second way?

To answer this question, we report in Table 2 the obtained statistical results using the average upper-level fitness value. It is important to note that evaluating the upper-level fitness corresponds to the measurement of the whole bi-level algorithm performance [9]. This point is explained by the fact that a BLOP consists in optimizing an upper-level objective function under some constraints, where one of these constraints corresponds to optimizing the lower-level problem (the follower objective function and constraints) (cf. Eq. (2)). Table 2 summarizes the average over all runs' evaluations using the number of FEs as a stopping criterion. As shown in this table, the best upper-level performance is given by our proposed CODBA based on Bi-MDVRPFD. Indeed, compared to the nested approach, our proposal generates the best upper-level fitness values for all test problems. In the nested approach, there is not any sophisticated mechanism that could either reduce the computational cost or accelerate convergence. As the evaluation of each upper-level solution needs running a whole lower-level process, thus to be able to generate good upper solution, the repair method needs to consume an extremely high number of FEs to achieve a promising upper-level result. Unfortunately, this is not efficient from a computational cost viewpoint especially with real application scenarios (high-dimension scenarios). For this reason, it has the worst performance in terms of upper-fitness values. For this reason, CODBA outperforms repair in terms of upper-level fitness values. The efficiency is ensured by the use of multiple parallel threads in lower level ensuring a saving regarding the classical nested scheme. Moreover, CODBA is designed using decomposition and co-evolution mechanisms helping for better exploration and convergence at lower level. This fact enhances the quality of upper solution (that depends on the lower reaction).

Table 2 Average upper-fitness values for CODBA and the nested algorithm on both Bi-MDVRPFD and Bi-MDVRP benchmarks

Bi-MDVRPFD				*Bi-MDVRP*		
INSTANCES		CODBA	REPAIR	INSTANCES	CODBA	REPAIR
Suite Bi-pr	**Bi-pr01-FD**	**2749 (+++)**	10360 (+++)	**Bi-pr01**	2926 (+++)	11296 (+++)
	Bi-pr02-FD	**5430 (+++)**	15989 (+++)	**Bi-pr02**	5954 (+++)	16068 (+++)
	Bi-pr03-FD	**12248 (-++)**	18049 (+++)	**Bi-pr03**	12339 (- ++)	19096 (+++)
	Bi-pr04-FD	17435 (+++)	26259 (+++)	**Bi-pr04**	**16816 (+++)**	27657 (+++)
	Bi-pr05-FD	**20660 (+++)**	30046 (++ -)	**Bi-pr05**	25636 (+++)	30325 (+-+)
	Bi-pr06-FD	32119 (+++)	33811 (+++)	**Bi-pr06**	**28969 (+++)**	37285 (+++)
	Bi-pr07-FD	**4703 (-++)**	19693 (+++)	**Bi-pr07**	4828 (+++)	22907 (+++)
	Bi-pr08-FD	**9961 (+++)**	26107 (+++)	**Bi-pr08**	11139 (+++)	22907 (+++)
	Bi-pr09-FD	**21261 (+++)**	36336 (+++)	**Bi-pr09**	22100 (+++)	38427 (+++)
	Bi-pr10-FD	**30227 (+++)**	46076 (+++)	**Bi-pr10**	31483 (+++)	53325 (+++)
Suite Bi-p	**Bi-p01-FD**	**1479 (+++)**	9529 (+++)	**Bi-p01**	1525 (+++)	10006 (+++)
	Bi-p02-FD	**1442 (+++)**	8880 (+++)	**Bi-p02**	1533 (+++)	12206 (+++)
	Bi-p03-FD	**2244 (+++)**	17676 (+++)	**Bi-p03**	2535 (+++)	14221 (+++)
	Bi-p04-FD	**3104 (-++)**	3932 (+++).	**Bi-p04**	3170 (+++)	6822 (+++)
	Bi-p05-FD	3284 (+++)	9674 (+++)	**Bi-p05**	3111 (+++)	10600 (+++)
	Bi-p06-FD	**3062 (- ++)**	7876 (+++)	**Bi-p06**	3233 (- ++)	8686 (+++)
	Bi-p07-FD	**3113 (+++)**	15733 (+++)	**Bi-p07**	3217 (+++)	13635 (+++)
	Bi-p08-FD	**35220 (+++)**	40251 (+++)	**Bi-p08**	54370 (+++)	40661 (+++)
	Bi-p09-FD	51381 (+++)	41030 (+++)	**Bi-p09**	**49705 (+++)**	43025 (+++)
	Bi-p10-FD	**50931 (+++)**	43706 (+++)	**Bi-p10**	54550 (+++)	48209 (+++)
	Bi-p11-FD	**50513 (+++)**	48306 (++ -)	**Bi-p11**	51358 (+++)	48458 (+-+)
	Bi-p12-FD	**5879 (-++)**	8583 (+++)	**Bi-p12**	5903 **(-++)**	14329 (+++)
	Bi-p13-FD	**5712 (+++)**	13184 (++ -)	**Bi-p13**	6303 (+++)	13060 (+-+)
	Bi-p14-FD	**5404 (+++)**	9911 (++ -)	**Bi-p14**	6199 (+++)	9809 (+-+)
	Bi-p15-FD	**20110 (-++)**	29278 (+++)	**Bi-p15**	20163 (-++)	31576 (+++)
	Bi-p16-FD	**20155 (+++)**	29227 (+++)	**Bi-p16**	20562(+++)	31026 (+++)
	Bi-p17-FD	**19078 (+++)**	29727 (+++)	**Bi-p17**	19619 (+++)	32258 (+++)
	Bi-p18-FD	**47584 (+++)**	64488 (+++)	**Bi-p18**	48291 (+++)	72220 (+++)
	Bi-p19-FD	**46405 (+++)**	63542 (+++)	**Bi-p19**	47277 (+++)	66522 (+++)
	Bi-p20-FD	46688 (+++)	59894 (+++)	**Bi-p20**	**46068 (+++)**	63721 (+++)
	Bi-p21-FD	**99730 (+++)**	128023 (+++)	**Bi-p21**	100364 (+++)	133550 (+++)
	Bi-p22-FD	**97578(+++)**	135668 (+++)	**Bi-p22**	100070 (+++)	130074 (+++)
	Bi-p23-FD	**98797(+++)**	130016 (++-)	**Bi-p23**	102897 (+++)	135166 (+++)

The symbol "+" means that H_0 is rejected, while the symbol "-" means the opposite. The best values are highlighted in bold

To evaluate now the effect of the flexibility depot feature of the new model, we analyze the obtained results of both CODBA and repair on Bi-MDVRPFD regarding Bi-MDVRP on the same table. We can observe from that table that the results are promising and the proposal algorithm not only increases the flexibility of transportation operation but also reduces total traveling cost.

6 Conclusion

In this chapter, we have introduced a new integrated formulation for the PD planning problem considering shared depots resources constraints as a specific case of the open VRP using the bi-level framework. Thus, two different decision-makers controlling the production and the distribution processes with a hierarchical relationship between them are suggested through the bi-level model. In order to solve the mixed integer proposed formulation, a cooperative decomposition bi-level algorithm was developed. Besides, 33 derived problem instances are considered in the statistical experimental study. The obtained results show that our proposed algorithm provides competitive results regarding the baseline repair method. Thus, it would be interesting to inject other constraint to both distribution and production levels and to evaluate the performance of CODBA on these problems. Indeed, we plan to design a cooperative behavior between the two levels to improve the quality of generated solutions. Another interesting research direction to propose a multi-objective variant of the model and solving it using well multi-objective evolutionary algorithms [31, 32].

Appendix A: CODBA (A CO-evolutionary Decomposition-Based Algorithm)

The idea and the principle working of CODBA are shown in Fig. 3.

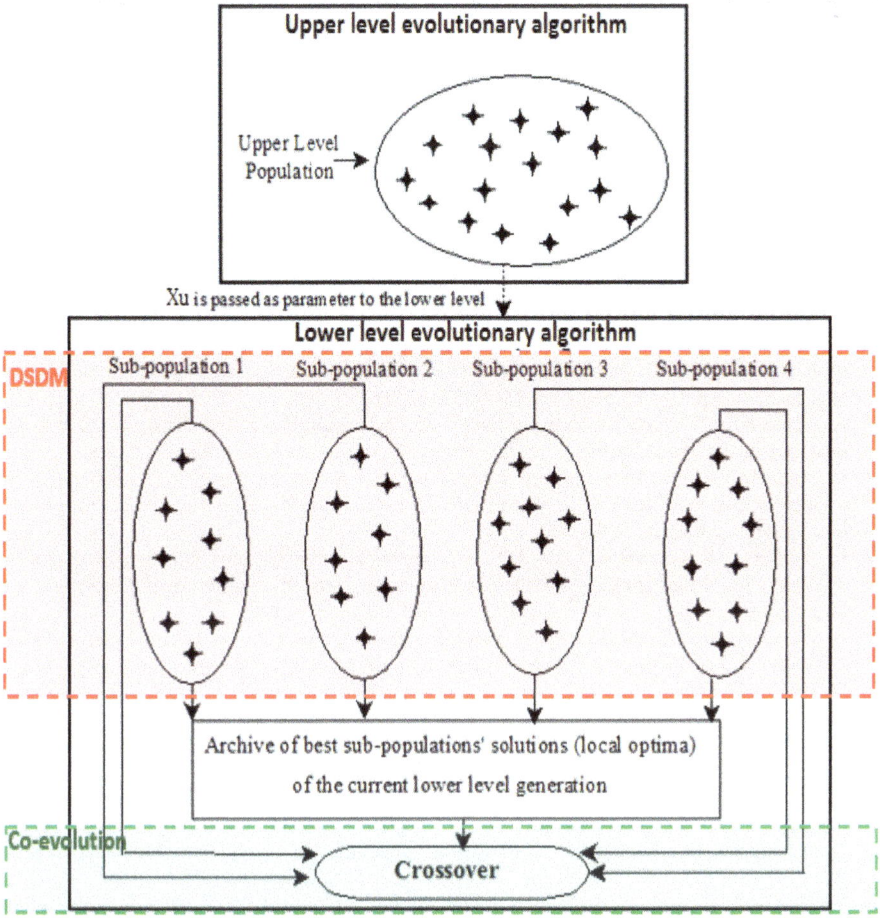

Fig. 3 Illustration of the CODBA scheme

References

1. R.G. Jeroslow, The polynomial hierarchy and a simple model for competitive analysis. Math. Program. **32**(2), 146–164 (1985)
2. H. Luo, L. Liu, X. Yang, Bi-level programming problem in the supply chain and its solution algorithm. Soft. Comput. **24**(4), 2703–2714 (2020)
3. H. Sun, Z. Gao, J. Wu, A bi-level programming model and solution algorithm for the location of logistics distribution centers. Appl. Math. Model. **32**(4), 610–616 (2008)
4. L.N. de Barros, W.R. Pinheiro, K.V. Delgado, Learning to program using hierarchical model-based debugging. Appl. Intell. **43**(3), 544–563 (2015)
5. P.A. Clark, A.W. Westerberg, Bilevel programming for steady-state chemical process design—I. Fundamentals and algorithms. Comput. Chem. Eng. **14**(1), 87–97 (1990)
6. M. Karaja, A. Chaabani, A. Azzouz, L. Ben Said, Efficient bi-level multi objective approach for budget-constrained dynamic bag-of-tasks scheduling problem in heterogeneous multi-cloud environment. Appl. Intell., **53**(8), 9009–9037 (2023)
7. A. Chaabani, L.B. Said, A co-evolutionary decomposition-based algorithm for the bi-level knapsack optimization problem. Int. J. Comput. Intell. Studies **9**(1-2), 52–67 (2020)
8. H.I. Calvete, C. Gale, A multiobjective bilevel program for production-distribution planning in a supply chain, in *Multiple Criteria Decision Making for Sustainable Energy and Transportation Systems* (2010), pp. 155–165
9. M. Abbassi, A. Chaabani, L.B. Said, An improved bi-level multiobjective evolutionary algorithm for the production-distribution planning system, in *International Conference on Modeling Decisions for Artificial Intelligence* (2020), pp. 218–229
10. A. Chaabani, L.B. Said, Transfer of learning with the co-evolutionary decomposition-based algorithm-II: a realization on the bi-level production-distribution planning system. Appl. Intell. **49**, 963–982 (2019)
11. A. Chaabani, S. Bechikh, L.B. Said, *A memetic evolutionary algorithm for bi-level combinatorial optimization: A realization between Bi-MDVRP and Bi-CVRP* (IEEE Congress on Evolutionary Computation (CEC), Vancouver, 2016), pp. 1666–1673
12. C.J. Vidal, M. Goetschalckx, Strategic production-distribution models: a critical review with emphasis on global supply chain models. Eur. J. Oper. Res. **98**(1), 1–18 (1997)
13. Ş.S. Erenguc, N.C. Simpson, A.J. Vakharia, Integrated production/distribution planning in supply chains: an invited review. Eur. J. Oper. Res. **115**(2), 219–236 (1999)
14. M. Mirabi, N. Shokri, A. Sadeghieh, Modeling and solving the multi-depot vehicle routing problem with time window by considering the flexible end depot in each route. Int. J. Supp. Oper. Manag. **3**(3), 1373–1390 (2016)
15. M. Rabbani, R. Moazemi, N. Manavizadeh, M.S. Jalali, Integrated supply, production, distribution planning in supply chain with regard to uncertain demand and flexibility in capacity, supply and delivery. Int. J. Math. Comput. Sci. **2**(1), 20–33 (2016)
16. Z. Moattar Husseini, B. Karimi, S.M. Moattar Husseini, S.H. Ghodsipour, Multi-objective integrated production distribution planning concerning manufacturing partners. Int. J. Comput. Integr. Manuf. **28**(12), 1313–1330 (2015)
17. Y. Liu, S. Li, K. Tang, Optimization models for integrated production-distribution planning with flexibility in depot resources, in *Proceedings of the International Conference on Industrial Engineering and Engineering Management (IEEM)* (2019)
18. D. Cao, M. Chen, Capacitated plant selection in a decentralized manufacturing environment: a bilevel optimization approach. Eur. J. Oper. Res. **169**(1), 97–110 (2006)
19. Y. Marinakis, M. Marinaki, A bilevel genetic algorithm for a real life location routing problem. Int. J. Logist. Res. Appl. **11**(1), 49–65 (2008)
20. H.I. Calvete, C. Gale, M.J. Oliveros, Bilevel model for production–distribution planning solved by using ant colony optimization. Comput. Oper. Res. **38**(1), 320–327 (2011)
21. J.-Y. Potvin, S. Bengio, The vehicle routing problem with time windows part II: genetic search. INFORMS J. Comput. **8**, 165–172 (1996)

22. K. Deb, M. Goyal, A combined genetic adaptive search (GeneAS) for engineering design. Comput. Sci. Inform. **26**, 30–45 (1996)
23. Y. Marinakis, A. Migdalas, P.M. Pardalos, A new bilevel formulation for the vehicle routing problem and a solution method using a genetic algorithm. J. Glob. Optim. **38**(4), 555–580 (2007)
24. J.F. Cordeau, G. Larporte, A. Mercier, A unified tabu search heuristic for vehicle routing problems with time windows. J. Oper. Res. Soc. **52**, 928–936 (2001)
25. J.F. Cordeau, M. Gendreau, G. Laporte, A tabu search heuristic for periodic and multi-depot vehicle routing problems. Networks **30**, 105–119 (1997)
26. N. Christofides, S. Eilon, An algorithm for the vehicle-dispatching problem. J. Oper. Res. Soc. **20**, 309–318 (1969)
27. B.E. Gillett, J.G. Johnson, Multi-terminal vehicle-dispatch algorithm. Omega **4**, 711–718 (1976)
28. I.-M. Chao, B.L. Golden, E. Wasil, A new heuristic for the multi-depot vehicle routing problem that improves upon best-known solutions. Am. J. Math. Manag. Sci. **13**, 371–406 (1993)
29. J. Li, Y. Li, P.M. Pardalos, Multi-depot vehicle routing problem with time windows under shared depot resources. J. Comb. Optim. **31**, 515–532 (2016)
30. J. Derrac, S. Garcıa, D. Molina, F. Herrera, A practical tutorial on the use of nonparametric statistical tests as a methodology for comparing evolutionary and swarm intelligence algorithms. Swarm Evol. Comput. **1**(1), 3–18 (2011)
31. M. Abbassi, A. Chaabani, N. Absi, L.B. Said, An elitist cooperative evolutionary bi-level multi-objective decomposition-based algorithm for sustainable supply chain. Int. J. Prod. Res. **60**(23), 7013–7032 (2022)
32. M. Abbassi, A. Chaabani, L.B. Said, N. Absi, Bi-level multi-objective combinatorial optimization using reference approximation of the lower level reaction. Procedia Comput. Sci. **176**, 2098–2107 (2022)

Analysis of Inhibitors to Implementing Digital Supply Chain in Saudi Arabia: An Interpretive Structural Modeling (ISM) Approach

Raouf Jaziri, Abdullah Alshareef, Saleh Alnahdi, and Mohammad Miralam

1 Introduction

The Internet boom and global business development have impressively metamorphosed customers' purchasing behavior and demand models, which apply strong pressure on supply chain managers. Consequently, the latter need to change their focus from reducing costs to enhancing new processes and making firms more agile to propose and create value to targeted markets and customers. During the past two decades, there has been a fundamental shift in the interaction tools between firms and customers who are becoming more exigent vis-à-vis channel response time and multichannel accessibility [1]. To overcome these challenges and expectations, firms should adopt the new framework of "digital supply chain" (DSC) as proposed by Farajpour et al. [2]. The recent systematic literature review carried out by many scholars [2–6] could be of great importance to identify the most suitable constructs for our current research. For research purposes, we retain the following definition provided by Büyüközkan and Göçer [7]: "*DSC (...) is an intelligent best-fit technological system that is based on the capability of massive data disposal and excellent cooperation and communication for digital hardware, software, and networks to support and synchronize interaction between*

R. Jaziri (✉)
College of Business, University of Jeddah, Jeddah, Saudi Arabia

Laboratoire d'Economie Dionysien, Université Paris 8, Paris, France
e-mail: rmjaziri@uj.edu.sa

A. Alshareef · S. Alnahdi · M. Miralam
College of Business, University of Jeddah, Jeddah, Saudi Arabia
e-mail: agalshareef@uj.edu.sa; saaalnahdi@uj.edu.sa; mmiralam@uj.edu.sa

I. Alharbi et al. (eds.), *Advances in Computational Logistics and Supply Chain Analytics*, Unsupervised and Semi-Supervised Learning,
https://doi.org/10.1007/978-3-031-50036-7_7

organizations by making services more valuable, accessible and affordable with consistent, agile and effective outcomes."

According to this definition, it is clear, that digital supply chain is an agile customer-oriented approach having as its main purpose to promote various types of returns and surplus for firms and to use in an efficient manner new digital technologies such as cloud computing, big data, and the Internet of things. Digitalization enables firms to shift from traditional supply chain to DSC by satisfying customers' needs and wants in a real-time perspective with a performant, agile, and responsive supply chain. Subsequently, this will lead to effective product accessibility with efficiency in terms of cost, lead, and delivery times [8]. Moreover, digital transformation is a strategic orientation that helps firms to upgrade their value chain toward innovative business models using digital technologies such as big business data analytics, cloud computing, and the Internet of things [9]. Supply chain digitalization allows firms to optimize their customer relationship management (CRM) and supplier relationship management (SRM) [10–12] and gives them new business and market opportunities in the road of new industrial revolution 4.0 [13, 14]. Despite the potential benefits of digital supply chain for industrial firms, its adoption remains a big challenge with probable resistance in Middle East countries [15]. As a result, the implementation of DSC conceals several enablers and barriers [16–24]. The literature review determines several barriers such as the unsuitable organizational structure, issue is not urgent, lack of industrial procedure manual, costly implementation, absence of digital skills, and absence of cyber security. This chapter discusses potential inhibitors to the implementation of DSC in Saudi Arabia. The choice of Saudi Arabia is justified because the kingdom's logistics performance index (LPI) has progressed in the last decade, and its current information and communication technology (ICT) infrastructure will accelerate the rhythm of logistical development. Moreover, given its strategic geographic location between three continents, Saudi Arabia reinforced its capability to become a global logistic hub in accordance with its prospective vision 2030. As a G20 economy and a crucial geopolitical actor in the region, Saudi Arabia had already launched huge investments in ICT to improve its supply chain management.

Inhibitors to the implementation of DSC in Saudi Arabia may have a mutual influence on each other. Therefore, the investigation of the mutual correlation between these inhibitors could be very interesting to guide Saudi policymakers in their decision-making process.

We strongly believe that the interpretive structural modeling (ISM) approach as a methodology for "*dealing with complex issues*" [25] is well suited to determine and analyze the different relationships between constructs [26–28]. Our current study determines inhibitors to DSC based on the available literature and opinions of Saudi academic and industrial experts. Then, the ISM approach is used to explore the interrelationships among inhibitors that could help decision-makers overcome these barriers when implementing DSC. This chapter has three main purposes:

1. Determine the essential inhibitors to supply chain digitalization.
2. Apply interpretive structural modeling (ISM) for the identified inhibitors.

3. Classify these inhibitors into four clusters using matrix of cross-impact multiplications applied to classification (MICMAC).

This chapter is structured into four main sections. Section 2 emphasizes the literature review associated with digital supply chain technologies and the different inhibitors to its implementation. Section 3 stresses the overview of the ISM approach as a selected methodology for modeling DSC's inhibitors. Subsequently, Sect. 3 classifies the inhibitors into four clusters using MICMAC analysis. Then, Sect. 4 discusses the obtained results. Finally, Sect. 5 concludes the chapter and underlines research limitations and future research orientations.

2 Literature Review

Business digital transformation necessitates a comprehension of customers' needs, wants, and their purchasing behavior during the product life cycle. DSC does not mean producing or selling digital products and services, but it is an approach to manage the main operations of supply chain [29, 30]. Moreover, the great Internet penetration rate has changed the customer's buying behavior and augmented the precariousness of demand. These challenges could be overwhelmed by using innovative technologies such as big data, cloud computing, and the Internet of things [31, 32]. Various challenges will be of great interest for firms especially internationalization, supply chain prominence and cooperation, modular product architecture, reactivity to unpredictable markets, and sustainable business model innovation [33, 34]. To deal with these challenges, firms across different industries are investing strongly in the digitization of their supply chains. Moreover, industrial firms are effectively increasing their digitization and are likely to be considered digital advanced companies. The industrial revolution reveals three crucial enablers especially satisfying unanticipated customer needs and wants, preserving product suppleness, and eventually realizing competitors' behaviors [35]. Thus, industrial revolution 4.0 refers to the use of artificial intelligence and innovative digital technologies to improve manufacturing processes and operation line [36].

Digital supply chain (DSC) could significantly create competitive advantage, increase customer satisfaction, strengthen partnership among supply chain stakeholders, reduce both delivery and lead times and manufacturing costs, and augment product availability [8, 37]. A large number of firms around the world have realized the importance of digitalization of their supply chain while there are a few firms that are satisfied with their uninterested attitude toward DSC [38, 39].

To respond to the research question of why firms resist the implementation of supply chain digitalization, Fitzgerald et al., have carried out a study among 1559 executives and managers in different American industries [40]. They identified that individuals are the main obstacle to digital transformation owing to their frustration and fear of unexpected results from innovative technologies. Thus, a recent literature review outlines the essential required abilities and methods for implementing a digital supply chain; although there is a lack of studies dealing with solutions to the organizational system's problems associated with digital organizational design and resource allocation.

The literature on DSC has been found using scientific databases such as Wiley, Springer, Taylor & Francis, Emerald, ScienceDirect, and Google Scholar. The literature chosen for the study consists of articles from reputable journals, industry reports, and book chapters. By a thorough analysis of the literature, 17 inhibitors to supply chain digitalization have been found (Table 1).

To make sure that these obstacles were appropriate for DSC in the Saudi context, a team of specialists was consulted. In this chapter, we identify 15 potential inhibitors to supply chain digitalization based on the recent literature review and the opinions of ten Saudi experts in supply chain management. These inhibitors are no urgent need, unsuitable organizational structure, lack of digital vision and strategic orientation, rigidity of business processes, biased business objectives, difficult adaptation to digital business transformation, risk aversion in taking initiative, lack of industrial accurate guidelines, great implementation cost, lack of top management support, lack of digital competencies, apprehension of cyber security risk, uncertain

Table 1 Inhibitors of DSC implementation

N°	Inhibitor	Author(s)
1	Lack of organizations' digital vision and strategy to succeed	Erol et al. [99]
2	Lack of financial resource	Mittal et al. [103]
3	Lack of governmental support and policies	Linh et al. [24]
4	Digital gap, constrain of technical resources, and usage of advanced manufacturing technologies (AMTs)	Mittal et al. [103] and Speece et al. [110]
5	Poor research and development on digital Supply Chain Management (SCM)	Schmidt et al. [108], Hermann et al. [100], and Mittal et al. [103]
6	Standard consideration for certification	Brown et al. [95]
7	Lack of cyber resources for supply chain	Radanliev et al. [106]
8	Cyber security and risk assessment	Anderson [94], Carruthers [96], and DiMase et al. [98]
9	Lack of standards and data sharing protocol	Linh et al. [24]
10	Legal issues	Schroder [109] and Muller et al. [105]
11	Profiling and complexity issues	Erol et al. [99] and Ras et al. [107]
12	Unclear economic benefit of digital investments	Kiel et al. [101] and Marques et al. [77]
13	Competition from multinational enterprises (MNEs) with better technological ability	Linh et al. [24]
14	Transaction risk	Mukhopadhyay and Kekre [104] and Clemons and Row [97]
15	Risk aversion	Linh et al. [24]
16	Human resource constraints	Kohnke [102]
17	Lack of coordination and partnership	Linh et al. [24]

return on digital investment, and fear of transaction risk and legal issues. A succinct description of these inhibitors is specified below.

3 Inhibitors to Digital Supply Chain in Saudi Arabia

3.1 No Urgent Need

The no urgent need for digitalization is one of the major obstacles to the implementation of DSC [40–42]. When an organization has a "status quo" mentality, it means that its current strategies satisfy its needs and that no change is required. According to Kotter, urgency is when many individuals have a strong desire to act for the organization in order to exploit a significant strategic opportunity [43].

Any digital transformation effort must start by creating and sustaining a logic of urgency. DSC could contribute to overcome the recent universal challenges; however, an acceptable level of organizational behavior and a strong organizational culture will be a significant inhibitor to digital transformation [41]. An individual who is satisfied with the current status quo sees no reason to change anything. To overpass this self-complacent behavior and to establish a relevant need of urgency, Kotter [43] proposes to "bring the outside in" and act on it. This could be attained by alleviating stakeholders' resistance toward perceived risks and emphasizing both the advantages and opportunities of implementing DSC.

3.2 Unsuitable Organizational Structure

The organizational framework conceived and created by managers to assign and coordinate human resources' activities is known as organizational structure. All employees are divided into groups according to their responsibilities and the adequacy between acquired and required skills as described by a system of hierarchy [44]. A strong organizational structure creates a hierarchy of reporting relationships and maintains open lines of communication among all employees. The organizational structure is indispensable for DSC, because it creates a communication channel between organization members and their environment. However, the existing organizational structure does not contribute to knowledge creation and sharing, as well as in terms of optimization of resource allocation. Moreover, lower-level employees are not satisfied with their jobs because they are deprived of their decision-making authority. All these factors hinder implementing DSC.

The way different supply chain partners interact, carry out their tasks, and make decisions to attain corporate goals is expressed by the organization's structure. It focuses on the organizations' external relationships and how the entire supply chain is interconnected [45]. Ishfaq et al. [46] argued that articulating a supply chain organizational structure should be in accordance with digital transformation awareness and with key success factors of implementing DSC.

DSC's organizational structure requires direct communication and data sharing between all supply chain stakeholders as opposed to the traditional organization needing indirect communication via intermediaries. More data synchronization and openness are provided by this kind of connectivity, ensuring that all decision-makers have access to the same data and enabling machines to make operational decisions.

A customer-driven, agile, more integrated, innovative, and adaptable organizational structure for firms engaged in supply chain digitalization will enable a more transparent flow of information throughout the organization, enhancing knowledge sharing, learning, collaboration, integration, and decision-making [47].

3.3 Lack of Digital Vision and Strategic Orientation

The strategic orientation of a firm aids in converting new concepts into a framework for digital transformation initiatives. It is crucial in determining the organization's overall goals and vision [48, 49].

An organization should choose different strategic orientations for a successful digital transformation endeavor, including customer orientation, technological orientation, competitor orientation, supplier orientation, and innovation orientation [50]. DSC should adopt a customer-centric strategy that encourages broad customization, proactive business, and intelligent automation. However, the conventional supply chain, which depends on stability and standardization, lacks these viewpoints. Strategic planning of DSC must involve two-way communication and take advantage of both top-down and bottom-up approaches [51–53].

3.4 Rigidity of Business Processes

When dealing with a digital business model, a product's design should be optimized for supply, manufacturing, and supply chain activities. Thus, products should be produced respecting the triad of cost, time, and place. Nowadays, business procedures reduce product life cycles and are unable to handle the difficulties of an expanding product portfolio [54]. Agile business procedures are necessary to satisfy a wide range of client wants and needs. The two most crucial characteristics of DSC are flexibility and a customer-oriented approach [55]. However, the existing procedures are not sufficiently flexible or responsive to consumer expectations [56]. Business processes for a DSC must be streamlined, flexible, and rapid to reconfigure [8]. They must be able to gather, analyze, and transform data into pertinent information that can be used in decision-making.

3.5 *Biased Business Objectives*

An obstacle to digitization is caused by differences between traditional and digital supply chain purposes. In contrast to traditional supply chain which is based on cutting costs and waste time reduction, the DSC is customer-driven, innovative, rapid, flexible, transparent, and scalable [7, 57].

3.6 *Difficult Adaptation with Digital Business Transformation*

Customers nowadays are increasingly demanding new digital goods and services, yet businesses' basic business processes and technological infrastructure cannot keep up with the dynamism and speed of the digital business environment. It necessitates a seismic and cultural shift in the work that people do, in their attitudes toward others, and in how they connect with others both inside and beyond the organization. The digital culture emphasizes more action and less planning, collaboration over individual work, and an external rather than internal orientation [58].

The ideals and distinctive set of behaviors that create an organization's culture define how things are done there [59]. It is crucial to evaluate the existing organizational culture in light of its digitization aspects features because if the current culture resists change, organizations may split into two parts: One part moves toward digitalization and the other part adheres to the traditional culture, which delays the transformation that is sorely needed. Managers must therefore take into account these cultural shifts that foster teamwork, worldwide connectedness, and consumer interactions in order to meet the dynamic of the digital business [60].

3.7 *Risk Aversion in Taking Initiative*

Digital transformation establishes new links inside the organization and with its partners, streamlines the operations, grows the businesses, and strengthens the customer relationships. The transformation requires high implementation and running costs, but there is risk associated with capturing transformation's return on investment (ROI) because there is a risk that organizational obstacles such as structure, culture, capabilities, and policies are not being adequately addressed to extract the full value of the transformation; there is a risk that leading team members will battle by evolving from their gradual and regular mode of working to a transformation mode, requiring a sense of urgency [61]. There is a lack of clarity about the pay-off from the investments in emerging technologies and the resistance from organizational and cultural factors that make it uncertain for employees to go for digital transformation.

3.8 Lack of Industrial Accurate Guidelines

It has been noted from the literature research that very few studies specifically focus on the DSC notion. Although there are numerous research papers, white papers, and industry reports that discuss digital technologies in terms of their applications in supply chains, very few of these papers explicitly focus on the adoption and implementation of DSC for a specific industry using case studies. Instead, they tend to focus on the technologies' enablers and anticipated benefits. Because businesses lack a clear understanding of which areas to convert first—internal operations, customer connections, or business models—a lack of industry-specific rules prevents the implementation of DSC. As a result, case studies on various industrial sectors should be carried out in order to provide a roadmap for the adoption of DSC in that particular industrial sector. Referring to the existing literature review, there are not many roadmaps for DSC implementation in that particular industrial sector. Many scholars argue that there are six best practices for industry 5.0 adopted by the European Commission [62, 63]. These six rules explicitly outline the conception of key indicators and metrics for digital measurement. The European goals EU2030 encourage industries in all European countries to digitalize their supply chain by exploiting opportunities related to innovative technologies [64, 65].

3.9 Great Implementation Cost

High implementation cost is considered the biggest obstacle to DSC [40, 41, 66]. To guarantee the availability of new digital technology, resources, a qualified workforce, and new organizational capacities, high investment is necessary. For their digital transformation efforts to be profitable, businesses must train their staff in relation to digital technologies. Technology systems are essential for DSC, but they need significant investment, and all of these tasks require funding. The significant implementation and operating costs of DSC must be balanced against its potential return on investment (ROI).

3.10 Lack of Top Management Support

According to many scholars, the lack of top management support is deemed as an inhibitor to the successful implementation of digital transformation in the supply chain [41, 67–69]. While the digitalization of the supply chain presents numerous opportunities, firms throughout the world are unable to take advantage of them due to a lack of top management support, effective leadership, and relevant experience. To give digital transformation programs a clear vision and value, top management support is necessary and imperative [70]. Without the support of senior management, employees will be resistant to change, especially older workers who are unaware of the benefits of the newest technology and do not want to learn about it. Without the involvement of top management, this mentality of the workforce cannot be changed [7]. The adoption of digital culture is also hampered by internal politics within the organization, which upper management can control. More than 20% of respondents to a survey agreed that internal politics, including the fear of losing control in the organization, impede the adoption of digital technologies [40]. Thus, top management should continue to support strategic planning for DSC implementation, which upper management has control over. More than 20% of respondents to a survey agreed that internal politics, including the fear of losing control in the organization, impede the adoption of digital technologies [40]. Thus, top management should continue to support strategic planning for DSC implementation.

3.11 Lack of Digital Competencies

The impact of new digital technology on the supply chain is enormous, but these effects cannot be fully realized without adequate human resources [71]. According to a study carried out by Hoberg et al., on digital talent, there is a serious lack of digital skills in the workforce [72]. The digital transition will be slowed down or delayed by a lack of digital talent. Workforce with a digital skill set, capable of leveraging digital technology and solutions and possessing durable business intelligence is required as a result of digitalization [73]. Organizations move quickly toward digitization once the DSC requirements have been met because this transformation depends on digital technologies and personnel. Businesses must evaluate their current skills and competencies before deciding which abilities will be crucial in the next years.

3.12 Apprehension of Cyber Security Risk

For businesses and individuals moving toward digitization, cyber security and privacy are the top concerns [74]. For DSC and its partners, security is and will continue to be a major concern. By 2018, the Internet will connect more than 13 billion devices and around 7 billion individuals and organizations, according to Gartner's hype cycle from 2014. Because of the extent of global interconnectedness, hackers will find it simple to cause significant disruptions and accomplish a large effect with modest inputs. A threat to the Internet "increasingly represents a threat to everything," according to the World Economic Forum in 2014.

According to the study carried out in 2015 by the Centre for Global Enterprise (CGE), 95% of respondents stated that the digitalization and sharing of business information with outside parties (suppliers and customers) had increased the need for cyber security measures. Digitalization will not only bring about enormous benefits, but it will also dramatically raise the hazards associated with cyber security.

3.13 Uncertain Return on Digital Investment

The use of smart technologies in conventional industries management can optimize the impact of the subsequent digitalization. Medvedeva [75] argued that the long-term return on digital investment represents 6.7 times the profit from investments in non-digital assets. Despite the potential effect of supply chain digitalization on increasing revenues and profits, its return on investment is still doubtful. One of the main obstacles to the adoption of technologies in the supply chain is the absence of economic and financial certainty about return on investment [76, 77]. Moreover, Chen and Lin argued that many organizations found problems in calculating the return on investment for various digital activities like using social media [78].

3.14 Fear of Transaction Risk

Yildiz et al., categorize three types of risks related to the adoption of digital supply chain systems, namely environmental risks, network-related risks, and organizational risks [79]. Firms are more concerned about both technological and transaction risks when implementing DSC systems [24]. The transaction risk related to the implementation of digital supply chain systems is less manageable [53]. It primarily refers to the unpredictability produced by self-interested actions taken by the partners in a transaction to misuse information resources and take advantage of the relationships [80].

Cybersecurity risks are increased when supply chains incorporate technology such as software-defined networks, storage, peer-to-peer connectivity, protocols, cloud computing, artificial intelligence, machine learning, and data analytics. Because adopting smart manufacturing technologies is typically expensive, SMEs may be forced to use less secure solutions. As a result, adding a cyber component to manufacturing also introduces a cyber risk [24, 53, 79].

3.15 Legal Issues

There are many intricate legal difficulties that arise in a developed cyber–physical network where many equipment, sensors, facilities, and people are connected to the Internet and share data with one another. Concerns regarding data privacy and security should be taken into account while implementing current technological management in business [81].

4 Research Methodology

The current study focuses on 15 inhibitors that are mutually exclusive. The relationships between the chosen inhibitors are determined, and the structural model is prepared using the interpretive structural modeling (ISM) technique.

The ISM is an "interpretive" approach because the relationships between variables are developed by expert judgment. It is also "structural" because the suggested relationships form a general structure derived from a complex group of variables. Moreover, ISM is a "modeling" approach since the explicit relationship and general structure are shown in a graphical commented model [82].

ISM has been widely used as a modeling tool in a variety of fields to scrutinize the influences and dependences between the numerous factors in an understudy system [83]. Managers and policymakers can visualize problems using ISM technique under a systematic approach. ISM can pinpoint the most influencing barriers that necessitate a crucial attention and determination to resolve them.

Due to the existence of several variables and their interactions, practitioners encounter numerous issues while working with complex systems. So, it is essential to employ an appropriate technique that forms a relationship between these obstacles, gives the variables rank and order, and lessens the complexity of the system. The aforementioned issue is resolved using the interactive decision modeling technique (ISM) [25, 84]. The system under study is affected by a variety of variables that are organized in a thorough core model by the ISM approach.

The specialty of this approach is that it makes it evident how important each piece that makes up the system is. It offers a ranking of the intricate relationships between a system's variables. By utilizing the ISM approach, Ullah and Narain [85] created a model that illustrates the hierarchical relationships between the various mass customization enablers. This work does not include a full explanation of ISM methodology; instead, readers should consult the literature [86].

Table 2 illustrates the key characteristics [87], advantages [88], and limitations [89].

5 The Process of ISM Methodology

ISM is an advanced planning methodology used to discover and characterize multiple causal linkages between several complex factors of a specific problem [90]. We use the MICMAC software to perform the structural analysis and to scrutinize the strongest potential inhibitors. The three different stages of the ISM process are described below.

Table 2 Characteristics, advantages, and limitations of ISM technique

Key characteristics	Advantages	Limitations
The method is interpretive since it establishes relationships between the various variables using expert judgment A digraph model is used to depict the related relationship, general structure, and analysis of direct and transitive linkages, making it a modeling tool It aids in simplifying the representation of complex systems	The revision and modification are easy The revision and modification are easy Require fewer computational exercises, when 10–15 variables are involved in the system The technique can be used for many real-life issues	The contextual relationship between the elements is highly dependent on the knowledge and experience of individuals ISM provides a feeble interpretation of links; therefore, other individuals cannot get a similar interpretation of the model

Table 3 List of variables

N°	Variable (long label)	Code (short label)
1	The need is not urgent	No_Urgency
2	Constrain of organizational structure	Org_Strctr
3	Rigidity of business processes	Bus_Proces
4	Lack of digital vision and strategic orientation	Dig_Vision
5	Biased business objectives	Bus_Objctv
6	Uncertain return on digital investment	RetDig_Inv
7	Fear of transaction risk	Trans_Risk
8	Risk aversion in taking initiative	Risk_Avers
9	Difficult adaptation with digital business transformation	Diff_Adapt
10	Lack of industrial accurate guidelines	Indst_Guid
11	Lack of top management support	Mng_Suport
12	Lack of digital competencies	Dig_Compet
13	Great implementation cost	Imple_Cost
14	Apprehension of cyber security risk	Secur_Risk
15	Legal issues	Legal_Issu

5.1 Stage 1: Listing Variables of the Understudied System

In the current study, the DSC's inhibitors are used as modeling factors. Fifteen barriers have been identified based on the current literature and the opinions of Saudi professionals and academia experts in supply chain management. Table 3 outlines the set of variables as listed in the input of MICMAC software.

Fig. 1 Matrix of direct influences (MDIs)

	1 : No_Urgency	2 : Org_Strctr	3 : Bus_Proces	4 : Dig_Vision	5 : Bus_Objctv	6 : RetDig_Inv	7 : Trans_Risk	8 : Risk_Avers	9 : Diff_Adapt	10 : Indst_Guid	11 : Mng_Suport	12 : Dig_Compet	13 : Imple_Cost	14 : Secur_Risk	15 : Legal_Issu
1 : No_Urgency	0	0	1	2	0	2	3	2	3	0	3	3	3	3	3
2 : Org_Strctr	0	0	2	2	0	0	0	3	2	2	1	0	0	0	0
3 : Bus_Proces	3	2	0	2	0	0	3	3	2	2	0	0	0	3	0
4 : Dig_Vision	3	3	3	0	3	3	3	3	3	3	3	3	1	3	2
5 : Bus_Objctv	0	0	0	3	0	0	0	0	1	0	0	0	0	0	0
6 : RetDig_Inv	3	0	0	2	0	0	3	3	3	0	3	2	3	3	0
7 : Trans_Risk	3	0	3	2	0	2	0	3	3	2	3	1	3	3	0
8 : Risk_Avers	0	2	2	1	0	0	3	0	2	0	2	3	0	3	1
9 : Diff_Adapt	3	2	1	1	0	2	3	2	0	1	3	2	3	0	0
10 : Indst_Guid	0	2	2	1	0	2	2	2	3	0	2	0	0	1	0
11 : Mng_Suport	3	2	3	3	2	3	3	3	3	2	0	2	3	3	2
12 : Dig_Compet	2	2	3	3	1	3	3	3	3	0	3	0	0	3	0
13 : Imple_Cost	3	0	0	3	3	3	2	0	2	1	3	2	0	2	0
14 : Secur_Risk	3	2	3	3	1	3	3	3	3	1	2	3	2	0	3
15 : Legal_Issu	3	3	3	3	0	1	3	3	3	2	3	2	0	3	0

5.2 *Stage 2: Description of the Relationship Between Variables*

After determining inhibitors, we interviewed a second group of ten industrial specialists who take the initiative to digitalize their supply chain management and we took into consideration their comments as part of the process of establishing the contextual relationship between each pair of variables. It was decided to use the "influence" relationship (for instance, inhibitor i influences inhibitor j). When using the software MICMAC, four conventions are used to fill the matrix of direct influences (MDIs) by indicating the direction of relationship between two inhibitors i and j (Fig. 1):

- "0" shows no influence from variable i to j.
- "1" shows low influence from I to j.
- "2" shows moderate influence from variable i to j.
- "3" shows strong influence from variable i to j.

5.3 *Stage 3: Identification of the Key Variables*

At this phase, the main variables, or those crucial to the development of the understudy system, are identified first using direct classification (which is simple to set up), then using indirect classification. After raising the matrix's power, this indirect classification is achieved.

6 Findings: MICMAC Analysis

There are two methods to analyze the interactions between variables included in the DSC system: direct relationship analysis and indirect relationship analysis using the matrix multiplication properties [91]. In direct relationship analysis, the direct relationships in the final ISM are considered to create a direct relationship matrix where the diagonal elements are set to zero and the transitive relationships are disregarded. The driving power of a barrier is calculated by adding the number of "1" entries in the row, and the dependency power of a barrier is calculated by adding the "1" entries in the columns. The ranks of the barriers are also estimated.

The direct influences matrix is schematized in Fig. 1. Although this reported matrix displays (MDI's) the greatest direct impact, it is unable to detect the variables' invisible impact, which is determined by their indirect association using the MICMAC method. The MICMAC analysis determines how dependent and powerful system components are, and it is dependent on the matrix multiplication characteristics [92]. Variables "i" and "k" would have an indirect relationship if variable "a" impacts variable "j," and variable "j" affects a third variable "k," In the system, there are many such indirect relationships that are not represented in the direct relationship matrix. MICMAC research shows a number of these indirect relationships that have an impact on the DSC as understudy system.

Based on the MDI, the direct influence/dependence map can be seen in Fig. 2.

Findings of this mapping outlined three inhibitors deemed as very influencing variables, namely Legal_Issu (legal issue), Imple_Cost (great implementation cost), and RetDig_Inv (uncertain return on digital investment). These three inhibitors were the triggers for the implementation of digital supply chain in Saudi Arabia.

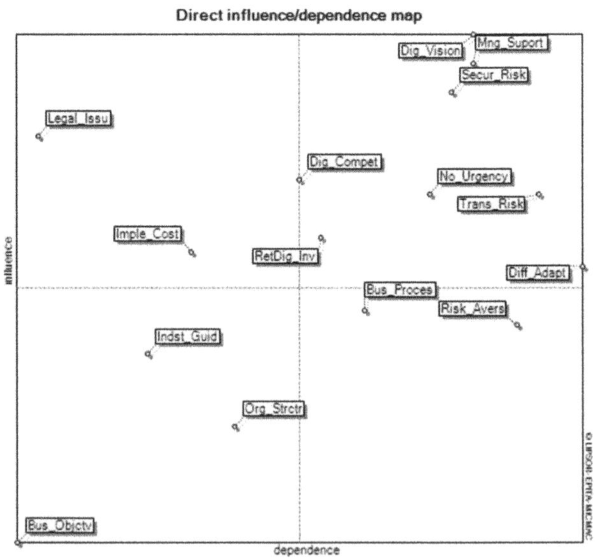

Fig. 2 Direct influence/dependence map

The implementation of DSC may be significantly inhibited by legal issues and regulations. In today's globalization era, supply chains are globally interconnected, and the movement of merchandises within them could engender a complex legal problem owing to the increased number of different stakeholders and rules [93]. The lack of a clear legislation and rules controlling the usage of DSC is also frustrating. Therefore, legal ambiguity could serve as an external impediment to the adoption of DSC.

One more critical inhibitor to implementing DSC, under the managerial perspective, is the high implementation cost [19, 41]. The implementation of DSC is accompanied by two crucial costs especially verification and network connectivity [93].

From an economic perspective, we consider the costs of implementing a digital supply chain as too high, while there is an uncertainty about the return on digital investment (RODI) and the payback time. Therefore, the interviewed experts consider the RODI to be very long and uncertain to justify the huge financial investment of the firm. This finding is in accordance with those of Hasanova and Romanovs [21], who suggest that when it comes to the process of digitalization of a traditional supply, we must be aware that firms make a significant initial investment.

The MICMAC approach can illustrate the impact of an inhibitor on another, as shown in the direct influence graph in Fig. 3. In this case, the "Legal_Issu" inhibitor has the strongest influences on six other inhibitors, namely rigidity of business processes, lack of digital vision and strategic orientation, fear of transaction risk, difficult adaptation to digital business transformation, lack of digital competencies, and apprehension of cyber security risk. On the other hand, the inhibitor "Risk_Avers" (risk aversion in taking initiative) is influenced especially by ten other inhibitors. Consequently, this demonstrates that all inhibitors of DSC have to be taken into consideration in accordance with their power of influence and dependence.

The indirect influence/dependency map (Fig. 4) outlines that the inhibitor "RetDig_Inv" is no longer an independent variable but is now included in the linkage variable, meaning that now this inhibitor is no longer a sensitive inhibitor but a stable one. The indirect relationship matrices are computed in the same way as done for the direct relationship matrices. The barriers are divided into four clusters according to the MICMAC analysis (Fig. 4). Every cluster includes a set of variables as detailed below:

1. *Cluster I—autonomous inhibitors*: They include obstacles with a lower driving and reliance power. These obstacles are not closely connected to the implementation of digital supply chain system. There are three inhibitors in this quadrant, namely Indus_Guid (lack of industrial accurate guidelines), Org_Strctr (constrain of organizational structure), and Bus_Objctv (biased business objectives).
2. *Cluster II—dependent inhibitors*: They include obstacles that have a substantial dependence but little driving force. By removing the restrictions that they rely on, these barriers can be removed as well. There are no inhibitors in this quadrant.

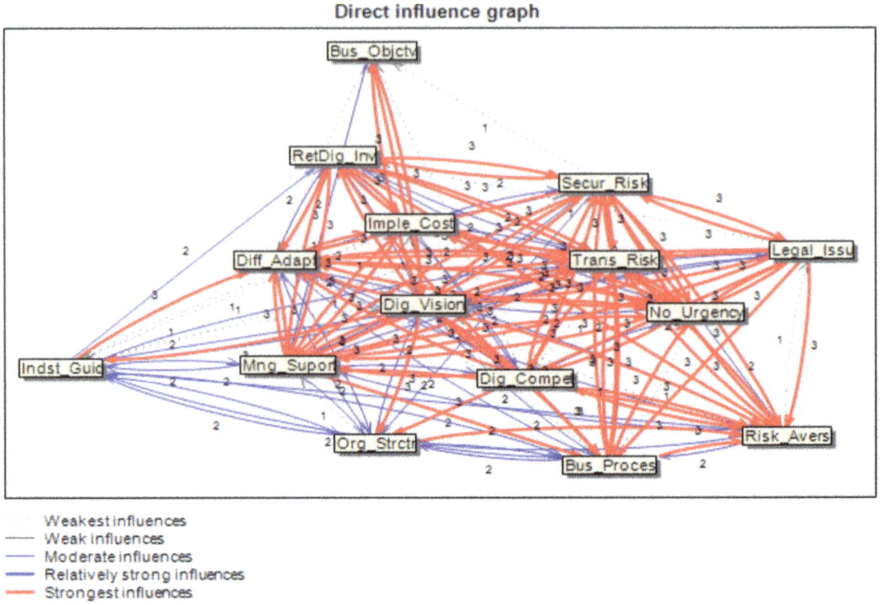

Fig. 3 Direct influence graph (scale 50%)

3. *Cluster III—linkage inhibitors*: They include obstacles that have both substantial influence and dependence forces. They frequently exhibit instability and reciprocal attachment with other obstacles. There are ten inhibitors in this quadrant especially lack of digital competencies (Dig_Compet), lack of digital vision and strategic orientation (Dig_Vision), uncertain return on digital investment (RetDig_Inv), lack of top management support (Mng_Suport), apprehension of cyber security risk (Secur_Risk), the need is not urgent (No_Urgency), fear of transaction risk (Trans_Risk), difficult adaptation with digital business transformation (Diff_Adapt), rigidity of business processes (Bus_Process), and risk aversion in taking initiative (Risk_Avers).
4. *Cluster IV—independent inhibitors*: They include obstacles that have a high driving force but a weak reliance. These inhibitors are considered the strongest barriers for implementing DSC in Saudi Arabia. Only two inhibitors are identified as independent and have a very powerful driving force, namely legal issues (Legal_Issu) and great implementation cost (Imple_Cost) (Figs. 4 and 5).

Meanwhile, the indirect influence graph shows that the inhibitor "Dig_Vision" (lack of digital vision) affects two other inhibitors, namely Diff_Adapt (difficult adaptation with digital business transformation) and Trans_Risk (fear of transaction risk). Moreover, the inhibitor "Mng_Suport" (lack of top management support) impacts two inhibitors indirectly; firstly, "Diff_Adapt" through "Risk_Aversion" and secondly, the "Trans_Risk" through "Legal_Issu." Therefore, the inhibitor "Secur_Risk" affects two other barriers directly, "Trans_Risk" and "Diff_Adapt."

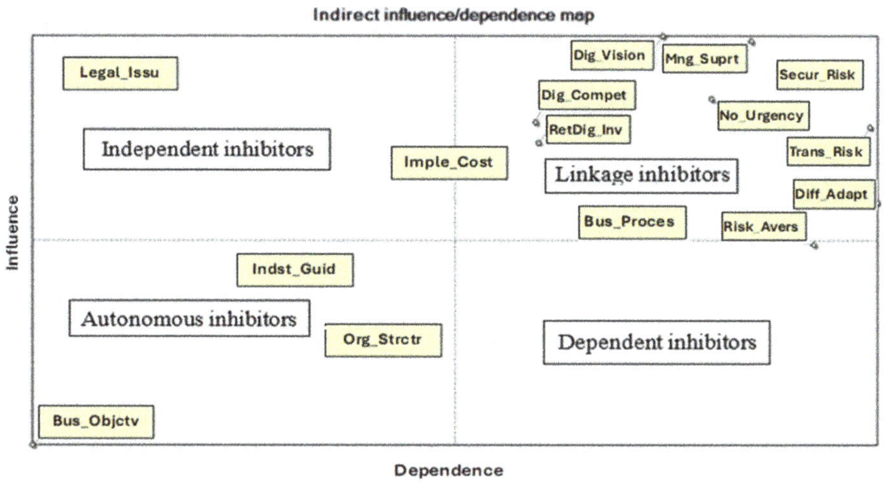

Fig. 4 Indirect influence/dependence map

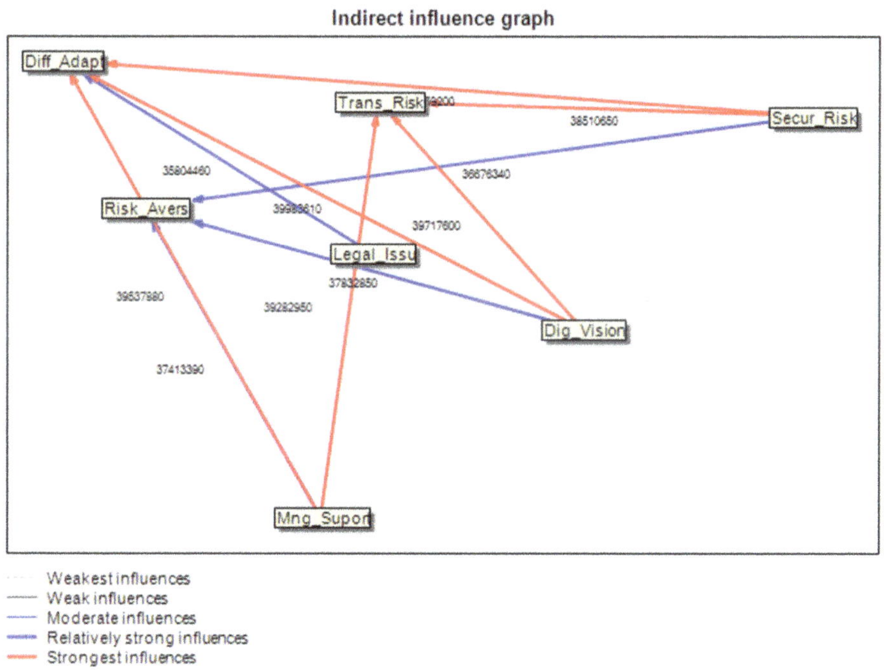

Fig. 5 Indirect influence/dependence graph (scale 25%)

The strongest influence is illustrated in the inhibitor "Dig_Vision" (lack of digital vision) as shown in Fig. 5.

7 Conclusions and Managerial Implications

It is evident from the findings above that a clear understanding of these inhibitors will automatically lead firms to recognize the significance of overcoming them. Moreover, firms will progressively take the required initiatives to reinforce their aptitudes to implement digital supply chain effectively.

The working environment within firms as well as with clients, suppliers, and partners has largely transformed by new digital technologies and apps. In these circumstances, firms must implement DSC. Since many firms have already begun integrating digital technology into their supply chains, other firms cannot afford to wait.

Management should foster an environment where innovation is strongly prioritized. It should be a commitment culture, and everyone in the firm should have a digital attitude. Firms must optimize the allocation of the available resources to encourage the use of DSC.

New abilities and skills are needed for digitization. Data scientists and analysts with the required skills have to gather, clean up, and analyze data. These professionals are essential for the supply chain's digital revolution. Other talents required for an organization's digital transformation include those in digital security, blockchain, mobile technology, social media, IoT, cloud computing, etc. At both the staff level and the CEO level, organizations' current workforces have significant skill gaps across all skill categories. Therefore, businesses need to make investments in specialized training and recruitment programs to strengthen the skill basis of their workforce for implementing cutting-edge digital technologies and solutions.

The process of digital transformation is complex, and each industry needs a specific roadmap to improve its DSC operations. It is still difficult to create a universally accepted process that consists of a series of steps for starting and managing digital transformation projects. As a result, firms cannot easily follow a digital transformation strategy.

The main purpose of this study is to list some of the most significant inhibitors stated in the DSC literature. Saudi companies can gain deeper understanding of the corresponding significance and interrelationships among the inhibitors using the ISM and the influence/dependence diagram offered in this study. As the field of DSC develops quickly, a few more obstacles can appear. Nonetheless, it emphasizes the necessity for a methodical approach to identifying the impediments, wisely allocating the resources at hand and establishing the required conditions for the adoption of DSC. Structural equation modeling can be used to further statistically validate the ISM given in this research.

The transition from conventional to DSC is happening quickly. Customers and competition are pressuring businesses to accelerate their digital transformation. All firms face pressure from customers and rivals to accelerate their digital transformation, which comes at a significant financial and labor expense. Yet, because resources must be transferred to enable this transformation, the speed of change is slow. This chapter outlines a methodical strategy for removing the obstacles that prevent the supply chain's digital transformation in the Saudi context.

References

1. T. Niranjan, P. Parthiban, Barrier analysis for channel synchronisation and implementation of omni channel supply chain strategy meeting customer preferences. Int. J. Six Sigma Competitive Advantage **14**(1), 138–151 (2022)
2. F. Farajpour, A. Hassanzadeh, S. Elahi, M. Ghazanfari, Digital supply chain blueprint via a systematic literature review. Technol. Forecast. Soc. Chang. **184**, 121976 (2022)
3. M. Aarland, T. Gjøsæter, Digital supply chain vulnerabilities in critical infrastructure: A systematic literature review on cybersecurity in the energy sector. ICISSP, 326–333 (2022)
4. R. Toorajipour, V. Sohrabpour, A. Nazarpour, P. Oghazi, M. Fischl, Artificial intelligence in supply chain management: A systematic literature review. J. Bus. Res. **122**, 502–517 (2021)
5. S. Kelly, V. Klézl, J. Israilidis, N. Malone, S. Butler, Digital supply chain management in the videogames industry: A systematic literature review. Comput. Games J. **10**(1), 19–40 (2021)
6. G.F. Frederico, From supply chain 4.0 to supply chain 5.0: Findings from a systematic literature review and research directions. Logistics **5**(3), 49 (2021)
7. G. Büyüközkan, F. Göçer, Digital supply chain: Literature review and a proposed framework for future research. Comput. Ind. **97**, 157–177 (2018)
8. J.R. Hanaysha, H.M. Alzoubi, The effect of digital supply chain on organizational performance: An empirical study in Malaysia manufacturing industry. Uncertain Supply Chain Manag. **10**(2), 495–510 (2022)
9. P.F. Anunciação, F.M. Esteves, Challenges to business models in the digital transformation context, in *Research Anthology on Digital Transformation, Organizational Change, and the Impact of Remote Work*, (IGI Global, 2021), pp. 116–129
10. G. Minculete, S.E. Stan, L. Ispas, I. Virca, L. Stanciu, M. Milandru, M.I. Bădilă, Relational approaches related to digital supply chain management consolidation. Sustainability **14**(17), 10727 (2022)
11. W. Cho, M. Fan, B. Yoo, H. Zhang, Special issue on digital transformation: Challenges and opportunities. IseB **19**(2), 387–388 (2021)
12. K.P. Liu, W. Chiu, Supply Chain 4.0: The impact of supply chain digitalization and integration on firm performance. Asian J. Bus. Ethics **10**(2), 371–389 (2021)
13. S.B. Ali, Industrial revolution 4.0 and supply chain digitization: Future of supply chain management. South Asian J. Social Rev. **1**(1), 21–41 (2022) ISSN: 2958-2490
14. M. Klippert, F. Marthaler, M. Spadinger, A. Albers, Industrie 4.0—An empirical and literature-based study how product development is influenced by the digital transformation. Proc. CIRP **91**, 80–86 (2020)
15. I. Manikas, B. Sundarakani, M. Shehabeldin, Big data utilisation and its effect on supply chain resilience in Emirati companies. Int J Log Res Appl, 1–25 (2022)
16. A. Kumar, V. Mani, V. Jain, H. Gupta, V.G. Venkatesh, Managing healthcare supply chain through artificial intelligence (AI): A study of critical success factors. Comput. Ind. Eng. **175**, 108815 (2023)
17. A.C. Benabdellah, K. Zekhnini, A. Cherrafi, J.A. Garza-Reyes, A. Kumar, J. Elbaz, Blockchain technology for viable circular digital supply chains: An integrated approach for evaluating the implementation barriers. Benchmarking (2022). https://repository.londonmet.ac.uk/id/eprint/8102
18. W.M.S.K. Weerabahu, P. Samaranayake, D. Nakandala, H. Hurriyet, Digital supply chain research trends: A systematic review and a maturity model for adoption. Benchmarking (2022). https://doi.org/10.1108/BIJ-12-2021-0782

19. M. Subramaniyam, S.A. Halim-Lim, S.F.B. Mohamad, A. Priyono, Digital supply chain in the food industry: Critical success factors and barriers, in *2021 IEEE International Conference on Industrial Engineering and Engineering Management (IEEM)*, (IEEE, 2021), pp. 404–410
20. M. Attaran, Digital technology enablers and their implications for supply chain management. Supply Chain Forum **21**(3), 158–172 (2020)
21. H. Hasanova, A. Romanovs, Best practices of technology management for sustainable digital supply chain, in *2020 61st International Scientific Conference on Information Technology and Management Science of Riga Technical University (ITMS)*, (IEEE, 2020), pp. 1–6
22. T. Mitra, R. Kapoor, N. Gupta, Studying key antecedents of disruptive technology adoption in the digital supply chain: An Indian perspective. Int. J. Emerg. Mark. (2022). https://doi.org/10.1108/IJOEM-07-2021-1052
23. G. Zhang, B.L. MacCarthy, D. Ivanov, The cloud, platforms, and digital twins—Enablers of the digital supply chain, in *The Digital Supply Chain*, (Elsevier, 2022), pp. 77–91
24. N.P.K. Linh, V. Kumar, X. Ruan, Exploring enablers, barriers and opportunities to digital supply chain management in Vietnamese manufacturing SMEs. Int. J. Organiz. Bus. Excel. **2**(2), 101–120 (2019)
25. F.R. Janes, Interpretive structural modelling: A methodology for structuring complex issues. Trans. Inst. Meas. Control. **10**(3), 145–154 (1988)
26. B.B. Gardas, R.D. Raut, B.E. Narkhede, A state-of the-art survey of interpretive structural modelling methodologies and applications. Int. J. Bus. Excell. **11**(4), 505–560 (2017)
27. N. Shin, S. Park, Evidence-based resilience management for supply chain sustainability: An interpretive structural modelling approach. Sustain. For. **11**(2), 484 (2019)
28. K. Zekhnini, A. Cherrafi, I. Bouhaddou, Y. Benghabrit, A. Belhadi, Supply Chain 4.0 risk management: An interpretive structural modelling approach. Int. J. Logist. Syst. Manag. **41**(1–2), 171–204 (2022a)
29. K. Zekhnini, A. Cherrafi, I. Bouhaddou, A. Chaouni Benabdellah, R. Raut, Barriers of blockchain technology adoption in viable digital supply chain, in *Product Lifecycle Management. Green and Blue Technologies to Support Smart and Sustainable Organizations: 18th IFIP WG 5.1 International Conference, PLM* (2022b)
30. A. Özdağoğlu, S. Bahar, Logistics 4.0 and smart supply chain management, in *Industry 4.0 and Global Businesses*, (Emerald Publishing Limited, 2022), pp. 163–183
31. B. Ageron, O. Bentahar, A. Gunasekaran, Digital supply chain: Challenges and future directions. Supply Chain Forum **21**(3), 133–138 (2020) Taylor & Francis
32. S. Bag, M.S. Rahman, G. Srivastava, A. Shore, P. Ram, Examining the role of virtue ethics and big data in enhancing viable, sustainable, and digital supply chain performance. Technol. Forecast. Soc. Chang. **186**, 122154 (2023)
33. S. Piqueras, Digital internationalizing firms (dif's): A systematic literature review and future research agenda. Piccola Impresa/Small Business **2** (2020)
34. P. Farahani, C. Meier, J. Wilke, Digital supply chain management agenda for the automotive supplier industry, in *Shaping the Digital Enterprise: Trends and Use Cases in Digital Innovation and Transformation*, (Springer, Cham, 2017), pp. 157–172
35. H. Gupta, S. Kumar, S. Kusi-Sarpong, C.J.C. Jabbour, M. Agyemang, Enablers to supply chain performance on the basis of digitization technologies. Ind. Manag. Data Syst. **121**(9), 1915–1938 (2021)
36. A. Shah, A. Ismail, Industrial revolution 4.0 and supply chain management, in *Integrating Blockchain Technology into the Circular Economy*, (IGI Global, 2022), pp. 82–92
37. M.M. Hussain Shahadat, A.H.M.Y. Chowdhury, R.J. Nathan, M. Fekete-Farkas, Digital technologies for firms' competitive advantage and improved supply chain performance. J. Risk Financ. Manag. **16**(2), 94 (2023)
38. J.E. Amenyawu, *Comprehensive Review of Digital Supply Chain Platforms Benefits to SMEs in Africa in Enhancing Their Performance in Africa in the Era of Digital Disruptions* (2021). Available at SSRN 3870602
39. A.F. AlMulhim, Smart supply chain and firm performance: The role of digital technologies. Bus. Process. Manag. J. **27**(5), 1353–1372 (2021)

40. M. Fitzgerald, N. Kruschwitz, D. Bonnet, M. Welch, Embracing digital technology: A new strategic imperative. MIT Sloan Manag. Rev. **55**(2), 1 (2014)
41. P. Agrawal, R. Narain, I. Ullah, Analysis of barriers in implementation of digital transformation of supply chain using interpretive structural modelling approach. J. Model. Manag. **15**(1), 297–317 (2020)
42. G. Salvini, G.J. Hofstede, C.N. Verdouw, K. Rijswijk, L. Klerkx, Enhancing digital transformation towards virtual supply chains: A simulation game for Dutch floriculture. Prod. Plan. Control **33**(13), 1252–1269 (2022)
43. J.P. Kotter, *Accelerate: Building Strategic Agility for a Faster-Moving World* (Harvard Business Review Press, 2014)
44. S. Modgil, R.K. Singh, S. Agrawal, Developing human capabilities for supply chains: An industry 5.0 perspective. Ann. Oper. Res., 1–31 (2023)
45. N. Janićijević, The mutual impact of organizational culture and structure. Econ. Ann. **58**(198), 35–60 (2013)
46. R. Ishfaq, B. Davis-Sramek, B. Gibson, Digital supply chains in omnichannel retail: A conceptual framework. J. Bus. Logist. **43**(2), 169–188 (2022)
47. R. Rajaguru, M.J. Matanda, Role of compatibility and supply chain process integration in facilitating supply chain capabilities and organizational performance. Supply Chain Manag. **24**(2), 315–330 (2019)
48. C. Heavin, D.J. Power, Challenges for digital transformation—Towards a conceptual decision support guide for managers. J. Decis. Syst. **27**(sup1), 38–45 (2018)
49. A. Abd-Rabo, S.A. Hashaikeh, The digital transformation revolution. Int. J. Hum. Educ. **3**(4), 124–128 (2021)
50. Q. Wang, Z. Wang, X. Zhao, Strategic orientations and mass customisation capability: The moderating effect of product life cycle. Int. J. Prod. Res. **53**(17), 5278–5295 (2015)
51. X. Zhu, S. Ge, N. Wang, Digital transformation: A systematic literature review. Comput. Ind. Eng. **162**, 107774 (2021)
52. A. Gurzawska, Towards responsible and sustainable supply chains–innovation, multi-stakeholder approach and governance. Philos. Manag. **19**(3), 267–295 (2020)
53. L. Xue, C. Zhang, H. Ling, X. Zhao, Risk mitigation in supply chain digitization: System modularity and information technology governance. J. Manag. Inf. Syst. **30**(1), 325–352 (2013)
54. I. Petryk, Restructuring of business processes for sustainability: Revealing the potential of reengineering and Kaizen. Law Bus. Sustain. Herald **1**(1), 33–42 (2021)
55. O.N. Denisova, O.V. Musina, Y.A. Tatarovsky, Digital transformation of business processes in a customer-oriented approach, in *Digital Technologies in the New Socio-Economic Reality*, (Springer, Cham, 2022), pp. 219–227
56. R. Ingleton, *The Digitisation of Everything* (Ernst and Young LLP, 2011). Available at: www.ey.com/Publication/vwLUAssets/The_digitisation_of_everything_-_How_organisations_must_adapt_to_changing_consumer_behaviour/%24FILE/EY_Digitisation_of_everything.pdf
57. C.R. Sahara, J.D.E. Paluluh, A.M. Aamer, Exploring the key factor categories for the digital supply chain. In *9th International Conference on Operations and Supply Chain Management* (2019), pp. 1–11
58. T. Hautala-Kankaanpää, The impact of digitalization on firm performance: Examining the role of digital culture and the effect of supply chain capability. Bus. Process. Manag. J. **28**(8), 90–109 (2022)
59. B.M. Tharp, *Defining Culture and Organizational Culture: From Anthropology to the Office* (Interpretation a Journal of Bible and Theology, Harworth, 2009). Available at: www.thercfgroup.com/files/resources/Defining-Culture-and-Organizationa-Culture_5.pdf
60. W.R. Ho, N. Tsolakis, T. Dawes, M. Dora, M. Kumar, A digital strategy development framework for supply chains. IEEE Trans. Eng. Manag. (2022)
61. D. Ivanov, A. Dolgui, B. Sokolov, The impact of digital technology and industry 4.0 on the ripple effect and supply chain risk analytics. Int. J. Prod. Res., 1–18 (2018). https://doi.org/10.1080/00207543.2018.1488086

62. M. Arunmozhi, V.G. Venkatesh, S. Arisian, Y. Shi, V.R. Sreedharan, Application of blockchain and smart contracts in autonomous vehicle supply chains: An experimental design. Transp. Res. E Logist. Transport. Rev. **165**, 102864 (2022)
63. S. Dixson-Decleve, P.A. Balland, F. Bria, K. Dunlop, E. Giovannini, D. Tataj, S. Schwaag Serger, *Industry 5.0: A Transformative Vision for Europe—ESIR Policy Brief No.3* (European Commission Directorate-General for Research Innovation, 2022). https://doi.org/10.2777/17322
64. A. Renda, S. Schwaag Serger, D. Tataj, A. Morlet, D. Isaksson, F. Martins, E. Giovannini, *Industry 5.0, a Transformative Vision for Europe: Governing Systemic Transformations Towards a Sustainable Industry* (European Commission, Directorate-General for Research and Innovation, 2022). https://doi.org/10.2777/17322
65. M. Breque, L. De Nul, A. Petridis, *Industry 5.0: Towards a Sustainable, Human-Centric and Resilient European Industry* (European Commission, Directorate-General for Research and Innovation, Publications Office, 2021), https://data.europa.eu/doi/10.2777/308407
66. A. Banerjee, F. Lücker, J.M. Ries, An empirical analysis of suppliers' trade-off behaviour in adopting digital supply chain financing solutions. Int. J. Oper. Prod. Manag. **41**(4), 313–335 (2021)
67. I. Vlachos, Implementation of an intelligent supply chain control tower: A socio-technical systems case study. Prod. Plan. Control, 1–17 (2021)
68. J. Al Suwaidi, R. Aydin, H. Rashid, *Investigating Barriers and Challenges to Artificial Intelligence (AI) Implementation in Logistic Operations: A Systematic Review of Literature* (2022)
69. J. Bughin, A. Holley, A. Mellbye, Cracking the digital code: McKinsey global survey results (2015). Available at: www.mckinsey.com/business-functions/business-technology/our-insights/cracking-the-digital-code
70. R. Johansen, *Leaders Make the Future: Ten New Leadership Skills for an Uncertain World* (Berrett-Koehler Publishers, 2012)
71. A. Dwivedi, S.K. Paul, A framework for digital supply chains in the era of circular economy: Implications on environmental sustainability. Bus. Strateg. Environ. **31**(4), 1249–1274 (2022)
72. P. Hoberg, H. Krcmar, G. Oswald, B. Welz, Skills for digital transformation. IDT Survey (2017). Available at: www.i17.in.tum.de/uploads/media/IDT-Survey_Report_2017_final.pdf
73. L.B. Liboni, L.O. Cezarino, C.J.C. Jabbour, B.G. Oliveira, N.O. Stefanelli, Smart industry and the pathways to HRM 4.0: Implications for SCM. Supply Chain Manag. **24**(1), 124–146 (2019)
74. C. Colicchia, A. Creazza, D.A. Menachof, Managing cyber and information risks in supply chains: Insights from an exploratory analysis. Supply Chain Manag. **24**(2), 215–240 (2019)
75. A.M. Medvedeva, Artificial intelligence as a new tool for growth of innovation and competitiveness of the digital business. Rev. ESPACIOS **40**(35) (2019) https://www.revistaespacios.com/a19v40n35/19403508.html
76. D. Kiel, J. Muller, C. Arnold, K.I. Voigt, Sustainable industrial value creation: Benefits and challenges of Industry 4.0, in *ISPIM Innovation Symposium*, (The International Society for Professional Innovation Management (ISPIM), 2017)
77. M. Marques, C. Agostinho, G. Zacharewicz, R. Jardim-Goncalves, Decentralized decision support for intelligent manufacturing in Industry 4.0. J. Ambient Intel. Smart Environ. **9**(3), 299–313 (2017)
78. S.C. Chen, C.P. Lin, Understanding the effect of social media marketing activities: The mediation of social identification, perceived value, and satisfaction. Technol. Forecast. Soc. Chang. **140**, 22–32 (2019)
79. S.C. Yildiz Ozenc, M. Er, S.U. Firat, Risks in supply chain 4.0: A literature review study. Chapter 13, in *Industrial Engineering in the Age of Business Intelligence*, ed. by F. Calisir, (Springer, Cham, 2023), pp. 133–177. https://doi.org/10.1007/978-3-031-08782-0_13
80. A. Chaouni Benabdellah, K. Zekhnini, A. Cherrafi, J.A. Garza-Reyes, A. Kumar, J. El Baz, Blockchain technology for viable circular digital supplychains: An integrated approach for evaluating the implementation barriers. Benchmarking (2023)

81. C. Schroder, *The Challenges of Industry 4.0 for Small and Medium-Sized Enterprises* (The Friedrich-Ebert-Stiftung, 2016)
82. M. Li, J. Yang, Analysis of interrelationships between critical waste factors in office building retrofit projects using interpretive structural modeling. Int. J. Constr. Manag. **14**(1), 15–27 (2014)
83. M.N. Faisal, B.A. Al-Esmael, Modeling the enablers of organizational commitment. Bus. Process. Manag. J. **20**(1), 25–46 (2014)
84. D.W. Malone, An introduction to the application of interpretive structural modelling. Proc. IEEE **63**(3), 397–404 (1975)
85. I. Ullah, R. Narain, Analysis of interactions among the enablers of mass customization: An interpretive structural modelling approach. J. Model. Manag. **13**(3), 626–645 (2018)
86. J. Thakkar, A. Kanda, S. Deshmukh, Supply chain issues in Indian manufacturing SMEs: Insights from six case studies. J. Manuf. Technol. Manag. **23**(5), 634–664 (2012)
87. D. Jain, D.K. Banwet, Modelling critical elements of selection for strategic alliance partner for network managed services using interpretive structural modelling (ISM). J. Model. Manag. **8**(3), 290–304 (2013)
88. T. Raj, R. Shankar, M. Suhaib, An ISM approach for modelling the enablers of flexible manufacturing system: The case for India. Int. J. Prod. Res. **46**(24), 6883–6912 (2008)
89. J. Jena, S. Sidharth, L.S. Thakur, D. Kumar Pathak, V.C. Pandey, Total interpretive structural modeling (TISM): Approach and application. J. Adv. Manag. Res. **14**(2), 162–181 (2017)
90. A. Sage, *Interpretive Structural Modelling: Methodology for Large-Scale Systems* (McGraw-Hill, New York, 1977), pp. 91–164
91. D. Jothimani, A.K. Bhadani, R. Shankar, Towards understanding the cynicism of social networking sites: An operations management perspective. Procedia Soc. Behav. Sci. **189**, 117–132 (2015)
92. D. Jothimani, R. Shankar, S.S. Yadav, Modeling hierarchical relationships among enablers of supply chain coordination in flexible environment, in *Managing Flexibility*, ed. by B.K.T. Sushil, S.P. Singh, (Springer, New Delhi, 2016), pp. 171–186
93. S. Joshi, A.A. Pise, M. Shrivastava, C. Revathy, H. Kumar, O. Alsetoohy, R. Akwafo, Adoption of blockchain technology for privacy and security in the context of industry 4.0. Wirel. Commun. Mob. Comput. **2022**, Article ID 4079781 (2022). https://doi.org/10.1155/2022/4079781. DEE
94. G. Anderson, The economic impact of technology infrastructure for Smart Manufacturing. NIST Econ. Anal. Briefs **4**, 1–6 (2016)
95. A. Brown, D.W.T. Van, K. Loughton, Smaller enterprises' experiences with ISO 9000. Int. J. Qual. Reliability Manag. **15**(3), 273–285 (1998)
96. K. Carruthers, Internet of things and beyond: Cyber-physical systems. IEEE Int. Things Newslett. **10** (2014)
97. E.K. Clemons, M.C. Row, Information technology and industrial cooperation: The changing economics of coordination and ownership. J. Manag. Inf. Syst. **9**(2), 928 (1992)
98. D. DiMase, Z.A. Collier, K. Heffner, I. Linkov, Systems engineering framework for cyber physical security and resilience. Environ. Syst. Decis. **35**(2), 291300 (2015)
99. S. Erol, A. Schumacher, W. Sihn, Strategic guidance towards Industry 4.0- A three stage process model, in *International Conference on Competitive Manufacturing* (2016)
100. M. Hermann, T. Pentek, B. Otto, Design principles for industry 4.0 scenarios, in *2016 49th Hawaii International Conference on System Sciences (HICSS)* (IEEE, 2016), pp. 3928–3937
101. D. Kiel, J.M. Müller, C. Arnold, K.I. Voigt, Sustainable industrial value creation: Benefits and challenges of industry 4.0. Int. J. Innov. Manag. **21**(08), 1740015 (2017)
102. O. Kohnke, It's not just about technology: The people side of digitization, in *Shaping the Digital Enterprise* (Springer, Cham, 2017), pp. 69–91
103. S. Mittal, M. Khan, D. Romero, T. Wuest, A critical review of smart manufacturing & Industry 4.0 maturity models: Implications for small and medium-sized enterprises (SMEs). J. Manuf. Syst. **49**, 194–214 (2018)

104. T. Mukhopadhyay, S. Kekre, Strategic and operational benefits of electronic integration in B2B procurement processes. Manag. Sci. **48**(10), 1301–1313 (2002)
105. P. Muller, H. Daniel, M. Sean, P. Viktoriya, H. Karen, K. Laura, J. Jenna, *Focus on Self-Employment. Annual Report on European SMEs 2016/2017* (European Commission, London, 2017)
106. P. Radanliev, D. Roure, J. Nurse, R. Montalvo, P. Burnap, Supply Chain design for the industrial internet of things and the Industry 4.0. SSRN Elect. J. (2019)
107. E. Ras, F. Wild, C. Stahl, A. Baudet, Bridging the skills gap of workers in industry 4.0 by human performance augmentation tools: challenges and roadmap, in *Proceedings of the 10th International Conference on Pervasive Technologies Related to Assistive Environments* (ACM, 2017), pp. 428–432
108. R. Schmidt, M. Mohring, H.C. Harting, C. Reichstein, P. Neumaier, P. Jozinovic, Industry 4.0 - Potentials for creating smart products: empirical research results, in *Proceedings of the International Conference on Business Information Systems* (Springer, 2015), pp. 16–27
109. C. Schröder, *The Challenges of Industry 4.0 for Small and Medium-Sized Enterprises*, vol. 7 (Friedrich-Ebert-Stiftung, Bonn, Germany, 2016), pp. 1–28
110. M. Speece, L. Tiangsoongnern, J. Roenjun, Digitization & globalization for the little players: The view from Thai SMEs, in *Proceedings of the 43rd Annual Macromarketing Conference* (2018), pp. 743–763

Financial Performance Measurement of Logistics Companies: Empirical Evidence from Saudi Arabia

Raéf Bahrini, Ahmed Zamzam, and Assaf Filfilan

1 Introduction

Supply chain management (SCM) involves planning and controlling all of the processes – from raw material production to purchase by the final customer – that link together all partners in a supply chain to meet the needs of the final customer [1]. An essential component of SCM is logistics, which is the task of managing the purchase and distribution of goods across the supply chain. Logistics is also defined as the process of managing the procurement, movement, and storage of materials, parts, and finished inventory across the organization and its marketing channels, with the aim of maximizing profitability through cost-effective order fulfillment [2]. Researchers and experts affirm that with globalization and integration of the world economy, logistics and SCM play an increasingly important role in ensuring a consistently high degree of customer satisfaction in terms of quality, delivery, and cost [3]. Without supply chains and logistics, products manufactured by suppliers cannot be efficiently distributed on global markets. Therefore, they are vital for markets all over the world [4].

In recent years, the global logistics industry has experienced remarkable growth, and logistics has become one of the key sectors of the business economic system and a major global economic activity. Logistics plays a huge role in the economies of most developing countries, influencing various areas, such as transport networks, storage systems, packaging services, exports and imports of services, information and communication devices, and so on. [5]. Due to the increasing distance between

R. Bahrini (✉) · A. Zamzam · A. Filfilan
College of Business, University of Jeddah, Jeddah, Saudi Arabia
e-mail: rbahrini@uj.edu.sa; afzamzam@uj.edu.sa; afilfilan@uj.edu.sa

I. Alharbi et al. (eds.), *Advances in Computational Logistics and Supply Chain Analytics*, Unsupervised and Semi-Supervised Learning,
https://doi.org/10.1007/978-3-031-50036-7_8

the points of procurement and consumption of goods on a global scale, the logistics industry has started to be considered an important tool for competition. For this reason, the logistics industry has become increasingly important in most countries around the world [6]. Many reports and research studies state that logistics activities accelerate economic growth and increase productivity. It has also been shown that efficient logistics is also important for the competitiveness of a country and could be a source of employment [7]. Thus, the development of the logistics sector plays a significant role, offering benefits in terms of growth and development; logistical investments change the functioning of a company and countries in general [8].

Supply chain and logistics play an essential role in the delivery of goods and services to end consumers. Given the importance of the logistics industry both at the economic level and at the country level, it is crucial for policymakers and managers to maintain and enhance the level of efficiency in the logistics sector. To achieve this goal, a performance measurement system is needed. It is therefore important for managers to determine the performance measures of logistics companies when providing their services. They need to develop appropriate performance measures and metrics to effectively manage logistics operations, which will improve the competitiveness of their companies. Despite the importance of identifying the relevant tools that will enable managers to effectively measure the performance of logistics companies during their logistics and supply chain processes, the literature review reveals that there is a limited number of studies on this topic [9]. Previous literature shows that managers need both financial and nonfinancial measures in the decision-making process. While senior managers need financial measures for management-level decisions, junior managers and workers need operational (nonfinancial) measures for conducting their daily business [10].

In this study, we focus mainly on the financial performance measures of logistics companies. Our purpose is to identify the best financial performance indicators for a sample of logistics companies listed on the Saudi Stock Exchange over the period 2018–2021. Company performance is evaluated using Shannon's entropy method, which is a multi-criteria decision-making (MCDM) method applied to choose the best option among different alternatives. We consider that this study will contribute to the literature by investigating the best financial performance measures in the case of Saudi logistics companies. In addition, we believe that this study will be useful for companies that trade in the logistic sector and will contribute to the literature as a reference for further studies [11].

The remainder of this study is structured as follows: Section 2 defines logistics and discusses the relationship between logistics and supply chain management (SCM). Section 3 focuses on the role and the growing importance of logistics in the global economy and provides an overview of the logistics industry in Saudi Arabia. In Sect. 4, we review the available literature on logistics and SCM measures. In Sect. 5, we discuss the methodology used. Section 6 presents the results and analysis. Concluding remarks, implications, and directions for future research are presented in Sect. 7.

2 Logistics and Supply Chain Management

2.1 *What Is Logistics?*

The concept of "logistics" has existed for centuries. Initially, the concept was used in military applications. Over time, and due to different periods of change, logistics has become more common. Many definitions of logistics have been found, and most of them refer specifically to material movement and storage. Waters [12] presents logistics as "the function responsible for the flow of materials from suppliers into an organization, through operations within the organization, and then out to customers." It entails organizing, carrying out, and managing commodities, services, and information from origin to consumption. Traffic, transportation, shipping, imports, exports, warehousing, inventory control, purchasing, and production planning are all coordinated by logistics. It is employed to plan, organize, and keep track of the resources required to move goods efficiently, effectively, and reliably.

Logistics can also be defined as the process of strategically managing the procurement, movement and storage of materials, parts and finished inventory (and the related information flows) in a manner that maximizes profitability through the cost-effective fulfillment of orders. From this viewpoint, logistics management is concerned with satisfying customer needs "through the coordination of materials and data streams that extend from the marketplace, through the firm and its operations and beyond that to suppliers" [13].

Due to intense international rivalry, businesses were compelled to offer cheaper prices, better quality, more lasting products, and greater product flexibility, particularly during the 1980s. Many programs, techniques, and technologies that were directly or indirectly related to logistics, such as just in time (JIT), helped for "inventory reductions and better coordination of the material flow along productive chains" in this changing context and due to the changing markets [14]. Companies began to look "beyond the box" as they realized the value and potential rewards of these cooperative partnerships with others in the upflow and downflow, suppliers, and clients.

2.2 *The Relationship Between Logistics and Supply Chain Management*

We have focused on the flow of materials through a single organization up to this point. However, organizations do not operate in a vacuum; rather, each one functions as a consumer when it purchases materials from its own suppliers and as a supplier when it distributes those goods to its own customers. For instance, a wholesaler operates as both a supplier and a customer when purchasing products from manufacturers to sell to retail stores. A supplier of components purchases raw materials from them, assembles them into components, and then sells the finished goods to other manufacturers. Most products move through a series of organizations as they travel between original suppliers and final customers.

Here, we focus more on the movement of materials, and then we will use the more general term "supply chain." A supply chain consists of a succession of activities and organizations that materials move through from beginning suppliers to ending consumers. Supply chain management (SCM) is the control of the entire workflow from raw materials to final goods. Food and health items, as well as things that enable us to work, travel, and have fun, would not be available to us without the supply chain. The supply chain is made up of a network of suppliers that are linked by a centralized management process. Each supplier serves as a link in a production chain that connects producers, merchants, and suppliers of raw materials [12].

SCM refers to a coordinated set of decisions and tasks aimed at integrating various stakeholders such as suppliers, manufacturers, warehouses, transporters, retailers, and customers to improve product efficiency, service availability, and distribution [15]. Moreover, Martin [13] explains that SCM, as the management of both upstream and downstream relationships with suppliers and customers, plays a vital role in providing high-quality customer value while minimizing costs across the entire supply chain. As a result, the focus of SCM is on relationship management to achieve a more profitable outcome for all parties in the chain. This poses significant challenges as there may be times when the narrow self-interest of one party must be sacrificed for the benefit of the entire chain.

Maia and Cerra [14] define SCM as the combination of essential procedures that control the movement of materials and information in both directions, inside the organization and among all the enterprises participating in the logistics network, until the ultimate end users receive the products. According to them, the main goal of SCM is to aggregate value for stakeholders and clients along these processes. Maia and Cerra [14] also state that "logistics is to be a constituting part of SCM," which oversees the flow of materials and information between companies in the same chain. It should not be forgotten that SCM refers to a range of processes focused on ensuring the smooth flow of goods and services from their source to final delivery. These processes include sourcing, procurement, outsourcing, partnership relations, product co-design, and more. However, there are important conditionings between logistics and SCM, and in practice, the relationship between the two fields is made quite complex because of the reciprocal effects of decision-making made in one field may have on the other.

Another difference is that logistics management focuses on the micro level of logistics, such as storage and warehousing, whereas SCM focuses on the macro level, such as logistics flow [16]. SCM refers to the complete set of activities that a company performs to manage the sourcing, procurement, conversion, and logistics of raw materials used in the manufacturing of their final products. The process involves close coordination and collaboration between all parties involved in the value chain, including suppliers, intermediaries, distributors, and customers, to ensure the timely delivery of high-quality products at the optimal price points. This complex system requires effective coordination and the use of logistics management tools to maintain efficiency and effectiveness. It links the initial supplier to the final consumer. In other words, logistics management is a subset of SCM that deals with the efficient management of goods.

3 The Role and Importance of Logistics

3.1 Role and Importance of Logistics in the Global Economy

The success of any organization is determined by the satisfaction of its customers. It is unlikely to survive in the long run if it fails to satisfy its customers. It will have great difficulty in making profits, generating high rates of return, adding shareholder value, or achieving any other measure of success. Consequently, organizations must provide products that satisfy customers. Unfortunately, customers evaluate products based on a variety of factors. Some of these factors are dependent on logistics, such as availability, which is determined by stock; delivery time, which is determined by transportation; damage, which is avoided through proper material handling; and the price, which is affected by logistics costs. So, the overarching goal of logistics can be expressed in terms of customer service. It must optimize material movement to attain high levels of client satisfaction.

Ultimately, the overall aim of logistics is to achieve high customer satisfaction. It must offer a high-quality service at a reasonable price [12]. By ensuring that goods are accessible at the appropriate time and location, logistics creates value. A product is said to have added place utility if it is available where it is needed, and time utility if it is delivered on schedule thanks to logistics. Martin [17] states that "logistics has always been a central and essential feature of all economic activity." Shapiro and Heskett [18] agree that there are few aspects of human activity that do not ultimately depend on the flow of goods from origin to consumption. Without logistics, materials cannot be moved, operations cannot be carried out, products cannot be delivered, and customers cannot be served.

Logistics is considered one of the important tools that contribute significantly to the change and improvement of economic indicators. At the macro level, logistics participates significantly in the national economy by creating jobs, generating national income, and attracting foreign investment. At the micro level, the logistics industry is critical to increasing the competitive power of corporations. In addition, the logistics industry has an important mission in revitalizing and improving the competitiveness of other industries.

Today, all industries are dependent on the logistics sector, which offers them important support. [6]. It is advantageous for both the logistics sector and the local economy to be interconnected. The quickly growing regional economic level can serve as a solid foundation for the growth of the region's logistics sector and significantly aid in advancing the upstream and downstream sectors of the logistics supply chain. Additionally, the growth of the logistics sector can help to advance the modification of the regional economic growth model, the reorganization of the industrial structure, the improvement of regional competitiveness, and the establishment of regional economic integration.

When Cao and Deng [19] examined the variables affecting the efficiency of the logistics sector in the Yangtze River economic belt, they found that industrial agglomeration, government intervention, and the sector's degree of openness to the outside world are all highly significant. The logistics sector's contribution to the Chinese economy was examined by Zhang and Zhao [20] using a logistic regression model, and they discovered that it has subsequently moved into a new stage of development. Using Jiangsu Province in China as an example, Cao and Rui [21] investigated the geographical evolution mode of the logistics industry agglomeration and discussed the extent to which factors such as the manufacturing sector and fixed investment had an impact on the logistics industry.

The logistics sector is primarily concentrated in urban regions, claim Kumar et al. [22]. The regional economy will be significantly impacted by the transportation agglomeration. To scrutinize the causal relationship between Pakistan's transportation infrastructure investment and long-term economic growth under the equilibrium model, Mohmand et al. [23] used the unit root cointegration analysis and the Granger causality test. They discovered a two-way causal relationship in economically developed provinces and a one-way causal relationship in less developed provinces. Tsekeris [24] examined how domestic logistics development affected Greek export trade and concluded that the expansion of the logistics sector has a sizable spatial spillover effect on local export trade. Regarding unidirectional influence, Kim et al. [25] looked at the connection between the logistics sector and the expansion of South Korean port cities' economies and concluded that hastening the growth of the port logistics sector can significantly increase employment rates and foster economic expansion. From the viewpoint of system dynamics, Liu [26] has created a model of dynamic coupling between the logistics sector and the local economy, contending that while the coordination of development in the short-term is challenging, harmonious growth in the medium- and long-term is possible.

As the global economy moves into the twenty-first century, logistics has become a crucial element of SCM and consumer demand. In less than two decades, logistics management has influenced the flow of products to meet or exceed consumer demand. Companies realized that by managing logistics, they could decrease costs and enhance productivity. By forming alliances with suppliers, shipping companies, and warehousers, and connecting these services via automated systems, the logistics of getting products to consumers is improved, resulting in lower overhead costs and faster delivery. Understanding how the logistics system theory works requires strategic planning when calculating what will be required while focusing on obtaining materials and managing the speed of product production to ensure timely delivery to the consumer. The construction of a workflow plan that minimizes costs by increasing visibility and improving the general understanding of business requirements is facilitated by streamlining communication and services between departments. Cost savings are produced by reduced warehousing expenses, supply forecast-based purchasing, better inventory control, dependable shipping, and on-time delivery to the customer.

3.2 *The Logistics Industry in Saudi Arabia: An Overview*

The foundation of Saudi Arabia's Vision 2030 program is to stimulate foreign investment in order to boost and diversify the country's economy and transform it into an international hub linking Asia, Europe, and Africa. This change will be greatly aided by its strategic geographic positioning, which is based on proximity to significant emerging markets and key marine routes. Government-led investments in the rail, maritime, road, airport, and logistics infrastructure have helped to grow and sustain the logistics sector. Due to its economic expansion, population aging, and accelerated urbanization, Saudi Arabia is investing in enhancing its transportation systems. This includes putting in place inter-urban networks such as freight and high-speed railways, as well as urban transportation systems such as metros and buses, through public–private partnerships with top global logistics businesses (public–private partnership (PPP)). In order to develop industrial clusters with multimodal freight ties to a variety of foreign destinations, the new strategy includes greater promotion of special economic zones (SEZs) in various regions of the nation. The Saudi government is committed to developing the sector and has invested substantial funds in future expansion plans. The combined sector of transport, communication, and storage accounted for approximately 6.6% of the gross domestic product (GDP) in 2020, which is evaluated as SR172.3 billion ($45.9 billion). It is probable that the aforementioned share will experience a continued ascent as a result of the execution of the Vision 2030 reforms, as well as the initiatives delineated in both the National Transformation Program (NTP) and the National Industrial and Logistics Development Program (NIDLP). According to current plans, the Saudi government will invest more than SR500 billion ($133.3 billion) in the expansion of ports, airports, rails, and various other infrastructure components through 2030 [27].

Saudi Arabia's goal in diversifying its economy is to strengthen the role of the private sector in transportation. Vision 2030 aims to have 20% of the ministry of transport projects self-financed, which opens up significant opportunities for private investment in ports, airports, rail, and road infrastructure [28]. Private companies are encouraged to work with the government to develop the Kingdom's transport infrastructure. Partnerships for operating seaports, airports, and the supply chains that support them are in high demand. Several major projects are being funded through public–private partnerships (PPPs), while several public transport facilities are preparing for full privatization.

Making the Kingdom a logistics hub capable of effectively integrating trade throughout Asia, Europe, and Africa is a crucial part of Saudi Arabia's vision 2030. The Kingdom has an edge over other countries thanks to its advantageous position, which might help it develop into a major regional logistics hub. The Saudi Arabia's logistics market is worth $18 billion. It is the third largest emerging market and accounts for 55% of the Gulf Cooperation Council (GCC) countries overall logistics business. With a value of around SR94 billion ($25 billion) in 2020, it is also one of the fastest growing logistics sectors in the world [28].

The Saudi government wants to enhance the country's capacity to receive Umrah and Hajj pilgrims from eight million to thirty million annually, improving Saudi Arabia's position in the world on the Logistics Performance Index from 49 to 25 [28]. To make room for market liberalization and private sector involvement, governance structures and laws, as well as import and export procedures, are being simplified. Additionally, it is believed that public–private partnerships will aid in financing infrastructure and offer expertise from leading logistics markets. By 2030, Saudi Arabia hopes to rank among the top logistics hub in the area.

Saudi Arabia has been able to decrease the time, expense, and variability of imports through process automation and reengineering. The amount of paperwork involved in import–export has decreased by 75%, and the average seaport clearance time has been lowered in half to 2.2 days [28]. It has become more predictable and dependable, with 40% of customs declarations in seaports now being completed within 24 h and 70% being cleared within 48 h. These results have been achieved by allowing declaration submission prior to arrival, digitizing declaration processing, making customs available 24 h a day, reducing the level of manual inspection by implementing enhanced risk management strategies and fostering improved collaboration, and integration among all government institutions involved in the import/export process. To eliminate infrastructure bottlenecks, Saudi Arabia is modernizing its airports and expanding its air cargo facilities. The goal is to increase the Kingdom's total air cargo capacity from 0.8 million tons per year today to 6 million tons per year in 2030.

Technology is enhancing security and control over the import–export process in Saudi Arabia. The status and development of shipments may now be followed by importers in real time. Customs brokers are asked to create their declarations as soon as the shipping manifest is made available online, that is, before the ship arrives. They receive automated updates on the progress of their shipments on their mobile devices. In order to provide effective and safe information sharing among all parties engaged in the import/export process, including vessel and terminal operations, electronic payment, and truck management, Saudi Arabia recently launched a port system. According to projections, the Saudi market for freight and logistics is poised to experience a compound annual growth rate (CAGR) of 6.53% by 2027. This segment is expected to grow due to projected increases in manufacturing activity, international trade, rising domestic consumer consumption, and the easing of government regulations. The cold chain segment of the Saudi logistics market has also grown in recent years. The active participation of the pharmaceutical industry and increasing demand for fresh/processed fruits, vegetables, meat, and dairy are expected to drive growth to $1.64 billion by 2020.

The first special economic zone (SEZ), which will be located in Riyadh, has been declared by the Saudi government and will help to foster future growth in the road freight industry. The SEZ will concentrate on integrated logistics, and those who reside there will enjoy unique benefits and norms designed to draw in additional global firms. As part of its Vision 2030, Saudi Arabia plans to encourage foreign investment and expand industry through the establishment of special zones with financial, trade, and visa exemptions. Integrated logistics will be prioritized at the new SEZ. All goods in zones with pending status, as well as those temporarily moved for maintenance or repair, will be exempt from the value-added tax (VAT).

4 Literature Review

Performance measures can be defined as means of quantifying the efficiency and effectiveness of an action. A metric is a term that refers to the definition of each performance measure, how it is calculated, and the data used in the calculation [29]. Business managers in all sectors need a consistent and well-designed performance measurement system to support them in the decision-making process aiming to link strategy to operations. This is particularly true for managers and decision-makers in logistics companies and supply chain systems [30–32].

A review of the literature shows that over the last few decades, numerous research studies have focused on the measurement of supply chain performance, which has become a topic of great interest to both researchers and professionals [33–35]. This is understandable, given the important role played by supply chain systems and logistics companies in today's markets. Indeed, effective logistics and supply chain management offers great opportunities to companies to improve their performance and become more competitive [36–38]. According to previous literature, supply chain performance can be measured using different approaches such as balanced scorecard approach (BSC), supply chain operations reference approach (SCOR), managerial approach, and process-based approach [9, 31, 33, 39–42].

As mentioned in the previous paragraph, researchers have used different methods and approaches to assess the performance of supply chain systems. Metrics used in these approaches are mainly financial and nonfinancial. Many earlier studies have suggested that companies should adopt a balanced approach when designing the performance measurement systems that they intend to use. This means that a better performance measurement system should include both financial and nonfinancial measures. Financial measures are needed to help managers make the right strategic decisions, while nonfinancial measures are essential for planning and controlling day-to-day operations such as production and delivery. Although it is recommended to see both financial and nonfinancial measures using balanced approaches, this does not mean neglecting any of them.

In their study, Galankashi and Rafiei [35] reviewed the recent research studies that have focused on measuring supply chain performance from a financial perspective. They show that the available literature on supply chain performance measurement, while very large and comprehensive, has in most cases neglected to emphasize the importance of financial measures. Many previous studies have not focused on studying the performance measures needed to evaluate and analyze the financial performance of supply chains and logistics companies [29].

Ritchie and Kolodinsky [43], Friemann et al. [44], and Hernaus et al. [45], among others, suggest that the financial performance of supply chains and logistics companies needs to be further studied in the literature. It is clear that many previous studies have focused primarily on the importance of moving from financial to nonfinancial performance evaluation of supply chains. However, these studies have neglected to provide an overview of the specific measures of financial performance applicable to supply chains and how they differ from other measures.

Friemann et al. [44] found that the measures used to evaluate the performance of a supply chain are often not directly linked to overall financial objectives. They suggest that the managers of logistics companies and supply chains should consider financial parameters when making decisions. It is therefore necessary to investigate the different measures of financial performance that could be useful in the case of supply chains and logistics companies. This could help us to know more about the best financial performance measures for supply chains and logistics companies.

According to Galankashi and Rafiei [35], the metrics most commonly used in the literature to measure supply chain financial performance are cost, sales, return on assets (ROA), asset turnover, inventory turnover, revenue growth, profit margin, return on investment (ROI), economic value added (EVA), market share, and cash-to-cash cycle time.

Erdoğan and Kırbaç [11] investigated the financial performance measures for a sample of Fortune 500 Turkish logistics companies between 2015 and 2019. Their methodology is based on multi-criteria decision-making (MCDM) techniques, which are the entropy method and the WASPAS method. They first use the entropy method to determine the best performance criteria affecting the financial performance of Turkish logistics companies; then the WASPAS method is used to rank these companies according to their performance levels. They also used a set of financial metrics, such as net sales, earnings before interest and taxes (EBIT), total assets, total equity, and export. They found that "export" is the best performance criteria affecting the performance of logistics companies in Turkey for all the years between 2015 and 2019.

5 Methodology and Data

5.1 Entropy Method

In this study, our purpose is to identify the best performance indicators affecting the financial performance of a sample of Saudi logistics companies listed on the Saudi Stock Exchange (Tadawul) over the period 2018–2021. To do this, we apply the entropy method, which is a multi-criteria decision-making (MCDM) method widely applied in previous literature. Using this method, we compute the weights of each decision criterion (financial ratio) in order to identify the best performance measures affecting the financial performance of the selected Saudi logistics companies.

The entropy method is a fixed weight method used to provide the amount of useful information based on evaluation criteria weights for the alternatives. It is applied to choose the best option among different alternatives by computing an entropy weight that measures the value of dispersion in decision-making. The greater the degree of dispersion, the greater the degree of differentiation, and more information can be derived [46, 47]. As suggested by Erdoğan and Kırbaç [11] and Lam et al. [48], the steps needed to determine the objective weight of each criterion based on the entropy method are summarized as follows:

Step 1. Construct a decision matrix, M: In the case of m alternatives and n evaluation criteria, the decision-making matrix, M is defined as follows:

$$M = \begin{pmatrix} x_{11} & x_{12} & \ldots & x_{1n} \\ x_{21} & x_{22} & \ldots & x_{2n} \\ \ldots & \ldots & \ldots & \ldots \\ x_{m1} & x_{m2} & \ldots & x_{mn} \end{pmatrix}$$

where x_{ij} shows the value of jth criterion from the ith alternative, where $i = 1,2,3,\ldots.,m$ and $j = 1,2,3,\ldots.,n$.

Step 2. Normalize the decision matrix, M: The decision matrix, M, is normalized for each criterion because of different scales of them, as in the following equation:

$$P_{ij} = \frac{x_{ij}}{\sum_{i=1}^{m} x_{ij}}, i = 1, 2, 3, \ldots.., m;\ j = 1, 2, 3, \ldots.., n$$

where P_{ij} is the normalized value.

Step 3. Determine the entropy value, e_j: After normalizing the decision matrix, D in step 2, we can measure the entropy value for all criteria. The entropy value is expressed as follows:

$$e_j = -k \sum_{i=1}^{m} P_{ij} \ln P_{ij}$$

where e_j is the entropy value of jth criterion. This value lies between zero and one, $0 \leq e_j \leq 1$. k is the entropy constant and P_{ij} is the normalized value.

Step 4. Determine the objective weight, w_j: The objective weight for each criterion is measured by the following equation:

$$w_j = \frac{1 - e_j}{\sum_{j=1}^{n} 1 - e_j}, j = 1, 2, 3, \ldots.., n$$

where w_j is the weight of the jth criterion. $0 \leq w_j \leq 1$ and $\sum_{j=1}^{n} w_j = 1$. A higher entropy weight (w_j) indicates that the evaluation criterion j is more important and useful for the decision-making process [49].

Table 1 Financial performance ratios used as evaluation criteria of Saudi logistics companies

Category	Financial ratios	Definition
Liquidity ratios	Current ratio (CR)	Current assets/current liabilities
	Quick ratio (QR)	(Current assets – inventories)/current liabilities
Solvency ratios	Debt-to-equity ratio (DER)	Total liabilities/equity
	Debt-to-assets ratio (DAR)	Total liabilities/total assets
Efficiency ratios	Asset turnover (AT)	Sales/total assets
Profitability ratios	Return on asset (ROA)	Net income/total assets
	Return on equity (ROE)	Net income/equity

Source: Self-constructed by the authors

5.2 Data

To apply the entropy method, we need to include data on m alternatives and n evaluation criteria. Alternatives here are Saudi logistics companies while evaluation criteria are financial performance ratios. Data used is related to ten selected logistics companies listed on the Saudi Stock Exchange (Tadawul). These companies belong to the following sectors: capital goods, commercial and professional services, transportation, and energy.

The financial ratios (evaluation criteria) required for the study were calculated from the financial statements of logistics companies available on the Saudi Stock Exchange website for all the years between 2018 and 2021. With reference to Galankashi and Rafiei [35], Erdoğan and Kırbaç [11], and Lam et al. [48], we include in our study a number of financial ratios as evaluation criteria. Table 1 shows the financial performance ratios used in this study, their definition, and their category.

The liquidity ratios measure a company's ability to meet its short-term obligations. The liquidity ratios used in the analysis are the current ratio (CR), which reflects the ability of companies to pay all outstanding claims and cover liabilities and the quick ratio (QR), which calculates the capacity of a firm to refund its current liabilities when payable with exclusively quick assets [50]. The solvency ratios consider the potential of companies to meet their total debts. The solvency ratios used in the analysis are the debt-to-equity ratio (DER), which measures the financial risk of a company by comparing the amount of its total debts with the value of its shareholders' equity and the debt-to-assets ratio (DAR), which measures the proportion of a company's assets financed by debt [51, 52]. The efficiency ratios are highly operational measures that analyze the effective use of resources to generate sales. One efficiency ratio is used in our analysis: the asset turnover (AT) ratio, which measures how well a firm's assets are used to generate sales [48, 53].

Table 1 shows two main profitability ratios, namely return on asset (ROA) and return on equity (ROE). ROA explicitly considers the assets used to support the company's activities. It determines whether the company is able to generate an adequate profit on its assets. ROE measures how much profit a company generates with the money invested by its shareholders [53, 54].

6 Results and Analysis

6.1 *Entropy Results*

By following the methodology previously described, the results are presented in the tables below. Table 2 displays the initial decision matrix for selected Saudi logistics companies in the period 2018–2021. It shows that our empirical analysis includes ten Saudi logistics companies and seven financial ratios. This means that we have selected m alternatives: $m = 10$ and determined n evaluation criteria, $n = 7$. We present the constructed initial decision matrix, M, in Table 2, as required in the first step of the entropy method. The initial decision matrix M is normalized, as required in the second step of the entropy method. The normalized decision matrix is shown in Table 3. As explained in the steps of the entropy method, we determine for each criterion the entropy value from the normalized decision matrix. Then, we compute the objective weight for each criterion. The results are reported in Table 4. It shows the entropy value (e_j) and the objective weight (w_j) for Saudi logistics companies in the years 2018–2021. According to Table 4, ROA and the ROE have the highest entropy weights in the years 2019, 2020, and 2021. This means that profitability ratios are the most affecting performance criteria for the Saudi logistics companies in the study period. However, there is one exception for 2018, in which DER was found to be the most influencing financial ratio. Table 4 also reveals that liquidity ratios (CR and QR) have moderately influenced the financial performance of Saudi logistics companies over the study period. Furthermore, efficiency ratios (AT) proved to be the least affecting performance measure between 2018 and 2021.

6.2 *Financial Performance Measurement Analysis Based on Entropy Results*

In the previous section, we presented the results obtained by applying the entropy method in the case of ten Saudi logistics companies over the period between 2018 and 2021. In this section, we will attempt to analyze these findings in order to identify the best financial measures that should be adopted by Saudi logistics companies.

We provide the ranking of financial performance ratios in Table 5 according to the results obtained by using the entropy method. Figure 1 also displays the entropy weights of the Saudi logistics companies' financial ratios on average over the period 2018–2021. As we can see from Table 5 and Fig. 1, ROE is ranked first because it has

Table 2 The initial decision matrix for Saudi logistics companies in the period 2018–2021

Companies	CR	QR	DER	DAR	AT	ROA	ROE
2021							
SISCO	2.0503	2.0016	1.1752	0.5403	0.1973	0.0190	0.0414
SGS	2.2935	2.2933	1.1401	0.5327	0.3318	−0.0525	−0.1123
SAPTCO	2.4422	2.4274	3.5555	0.7805	0.2009	−0.0461	−0.2101
BATIC	0.9813	0.9538	3.1221	0.7574	0.4881	−0.0175	−0.0723
BUDGET SAUDI	1.4793	1.4400	0.1788	0.1517	0.5244	0.1167	0.1376
THEEB	0.4708	0.4612	1.6758	0.6263	0.4956	0.0828	0.2216
BAHRI	0.5839	0.5119	1.1737	0.5399	0.2472	0.0108	0.0235
ALDREES	0.6337	0.5677	4.4619	0.8169	1.6760	0.0325	0.1777
SADR	4.3747	4.0639	0.3395	0.2534	0.2990	0.0147	0.0197
SIECO	1.9702	1.9702	0.8632	0.4633	0.6483	−0.6163	−1.1482
2020							
SISCO	0.9497	0.9125	1.5779	0.6121	0.2316	0.0492	0.0492
SGS	3.5026	3.5024	0.7674	0.4342	0.2834	−0.1029	−0.1818
SAPTCO	1.3490	1.3280	2.6852	0.7286	0.2490	−0.0804	−0.2963
BATIC	2.0974	2.0244	1.2219	0.5499	0.6031	−0.0162	−0.0361
BUDGET SAUDI	2.4561	2.1955	0.1650	0.1416	0.6581	0.1265	0.1474
THEEB	0.4534	0.4348	1.7152	0.6317	0.5048	0.0484	0.1313
BAHRI	1.6473	1.5260	1.0103	0.5026	0.3977	0.0765	0.1538
ALDREES	0.5787	0.5245	4.0004	0.8000	1.1146	0.0271	0.1357
SADR	2.3359	1.5086	0.9643	0.4909	0.8046	0.0382	0.0751
SIECO	2.9416	2.6623	0.4838	0.3261	0.7921	−0.3203	−0.4753
2019							
SISCO	0.9862	0.9440	0.8265	0.4525	0.2326	0.0268	0.0489
SGS	3.3481	3.3477	0.4675	0.3186	0.5875	0.0979	0.1437
SAPTCO	0.8454	0.8149	1.4972	0.5996	0.4180	0.0083	0.0207
BATIC	2.5862	2.4896	0.4505	0.3106	0.8241	0.0122	0.0176
BUDGET SAUDI	0.8837	0.8112	0.2017	0.2017	0.7183	0.1229	0.1540
THEEB	0.4382	0.4234	1.8432	0.6483	0.4948	0.0910	0.2586
BAHRI	1.5493	1.3734	1.1247	0.5294	0.3194	0.0313	0.0665
ALDREES	0.6330	0.5707	3.0985	0.7560	1.5846	0.0814	0.3335
SADR	2.4404	0.9052	0.2516	0.2010	1.0263	−0.0464	−0.0582
SIECO	6.8473	6.5539	0.1651	0.1417	1.2762	−0.2003	−0.2329
2018	1.0185	0.9507	0.7040	0.4131	0.2109	0.0245	0.0417
SISCO	3.3574	3.3571	0.3691	0.2695	0.6485	0.0935	0.1281
SGS	1.0724	1.0177	0.9509	0.4874	0.4684	0.0074	0.0144
SAPTCO	3.0069	2.8477	0.2988	0.2301	1.0117	0.0313	0.0406
BATIC	0.9035	0.8953	0.2752	0.2158	0.7247	0.1183	0.1509

(continued)

Table 2 (continued)

Companies	CR	QR	DER	DAR	AT	ROA	ROE
BUDGET SAUDI	0.3745	0.3551	1.3250	0.5699	0.6026	0.0683	0.1589
THEEB	1.6027	1.3639	1.1588	0.5368	0.2892	0.0230	0.0497
BAHRI	0.5382	0.4864	4.3346	0.8125	1.5085	0.0195	0.1040
ALDREES	3.4754	1.1924	0.1627	0.1399	0.9287	0.0108	0.0125
SADR	7.5092	7.4340	0.1373	0.1207	1.5416	0.0837	0.0952
SIECO	1.0185	0.9507	0.7040	0.4131	0.2109	0.0245	0.0417

Source: Authors' calculations based on financial data provided by the Saudi Stock Exchange website. Notes: Abbreviations are used for logistics company names, as follows: Saudi Industrial Export Co. (SIECO), Sadr Logistics Co. (SADR), National Shipping Company of Saudi Arabia (BAHRI), Aldrees Petroleum and Transport Services Co. (ALDREES), Saudi Industrial Services Co. (SISCO), Saudi Ground Services Co. (SGS), Saudi Public Transport Co. (SAPTCO), Batic Investments and Logistics Co. (BATIC), United International Transportation Co. (BUDGET SAUDI) and Theeb Rent a Car Co. (THEEB). We provide the values of financial ratios for each year between 2018 and 2021

the highest entropy weight obtained from the entropy method, which is 0.3257. This means that ROE is determined as the best financial performance criterion affecting the Saudi logistics companies' performance on average over the period 2018–2021. In addition, Table 5 and Fig. 1 show that ROA is ranked second. The weight value of ROA is high (0.3120) and very close to the weight value of ROE. It is clear here that the profitability ratios (ROE and ROE) are found to be the best financial performance criteria for Saudi logistics companies in the period of study.

Furthermore, Table 5 and Fig. 1 reveal that DER is ranked third with an average entropy weight of 0.1122 over the period 2018–2021. This result indicates how this solvency ratio is important for the financial performance of Saudi logistics companies. However, it is demonstrated from the results presented in Table 5 that CR and QR are ranked in the fourth and fifth position, respectively, which means that they are significant, but their effect on the financial performance of Saudi logistics companies is moderate. In addition, Table 5 indicates that AT and DAR have the lowest entropy weights on average, and then they are the least significant financial performance indicators in the case of Saudi logistics companies during the period of study.

According to Lam et al. [48], it is important to analyze the financial performance of logistic companies by considering the categories of financial ratios. In this study, we considered four broad categories of financial ratios, namely liquidity ratios, solvency ratios, efficiency ratios, and profitability ratios. Table 6 shows the entropy weights for each category of financial ratios between 2018 and 2021. This allows us to rank each category according to its entropy weight values. Figure 2 shows the entropy weights of financial ratios classified by category over the period 2018–2021 for the selected Saudi logistics companies. Based on the category weights obtained by applying the entropy method, Table 6 and Fig. 2 confirm the previous analysis that the most significant category of financial performance measures in the case of Saudi logistics companies is profitability ratios followed by liquidity and solvency

Table 3 The normalized decision matrix for Saudi logistics companies in the period 2018–2021

Companies	CR	QR	DER	DAR	AT	ROA	ROE
2021							
SISCO	0.1187	0.1199	0.0664	0.0989	0.0386	−0.0418	−0.0449
SGS	0.1327	0.1374	0.0645	0.0975	0.0649	0.1152	0.1219
SAPTCO	0.1413	0.1454	0.2010	0.1429	0.0393	0.1012	0.2280
BATIC	0.0568	0.0571	0.1765	0.1387	0.0955	0.0385	0.0785
BUDGET SAUDI	0.0856	0.0863	0.0101	0.0278	0.1027	−0.2560	−0.1493
THEEB	0.0272	0.0276	0.0948	0.1147	0.0970	−0.1817	−0.2405
BAHRI	0.0338	0.0307	0.0664	0.0988	0.0484	−0.0237	−0.0255
ALDREES	0.0367	0.0340	0.2523	0.1496	0.3281	−0.0714	−0.1928
SADR	0.2532	0.2435	0.0192	0.0464	0.0585	−0.0322	−0.0214
SIECO	0.1140	0.1180	0.0488	0.0848	0.1269	1.3520	1.2460
2020							
SISCO	0.0519	0.0549	0.1081	0.1173	0.0411	−0.3195	−0.1656
SGS	0.1913	0.2107	0.0526	0.0832	0.0503	0.6684	0.6122
SAPTCO	0.0737	0.0799	0.1840	0.1396	0.0442	0.5226	0.9979
BATIC	0.1145	0.1218	0.0837	0.1054	0.1070	0.1055	0.1214
BUDGET SAUDI	0.1341	0.1321	0.0113	0.0271	0.1167	−0.8222	−0.4963
THEEB	0.0248	0.0262	0.1175	0.1211	0.0895	−0.3144	−0.4423
BAHRI	0.0900	0.0918	0.0692	0.0963	0.0705	−0.4973	−0.5180
ALDREES	0.0316	0.0316	0.2742	0.1533	0.1977	−0.1763	−0.4569
SADR	0.1276	0.0908	0.0661	0.0941	0.1427	−0.2484	−0.2528
SIECO	0.1606	0.1602	0.0332	0.0625	0.1405	2.0817	1.6005
2019							
SISCO	0.0480	0.0518	0.0833	0.1088	0.0311	0.1191	0.0650
SGS	0.1629	0.1836	0.0471	0.0766	0.0785	0.4351	0.1910
SAPTCO	0.0411	0.0447	0.1508	0.1441	0.0559	0.0368	0.0275
BATIC	0.1258	0.1365	0.0454	0.0747	0.1101	0.0540	0.0234
BUDGET SAUDI	0.0430	0.0445	0.0203	0.0485	0.0960	0.5463	0.2047
THEEB	0.0213	0.0232	0.1857	0.1559	0.0661	0.4042	0.3437
BAHRI	0.0754	0.0753	0.1133	0.1273	0.0427	0.1391	0.0884
ALDREES	0.0308	0.0313	0.3121	0.1818	0.2118	0.3616	0.4432
SADR	0.1187	0.0496	0.0253	0.0483	0.1372	−0.2062	−0.0773
SIECO	0.3331	0.3594	0.0166	0.0341	0.1706	−0.8900	−0.3095
2018							
SISCO	0.0446	0.0478	0.0725	0.1088	0.0266	0.0509	0.0524
SGS	0.1469	0.1687	0.0380	0.0710	0.0817	0.1947	0.1609
SAPTCO	0.0469	0.0511	0.0979	0.1284	0.0590	0.0154	0.0181
BATIC	0.1315	0.1431	0.0308	0.0606	0.1275	0.0651	0.0510
BUDGET SAUDI	0.0395	0.0450	0.0283	0.0569	0.0913	0.2464	0.1896
THEEB	0.0164	0.0178	0.1364	0.1501	0.0759	0.1422	0.1996
BAHRI	0.0701	0.0685	0.1193	0.1414	0.0364	0.0480	0.0625
ALDREES	0.0235	0.0244	0.4461	0.2141	0.1901	0.0406	0.1307
SADR	0.1520	0.0599	0.0167	0.0369	0.1170	0.0225	0.0157
SIECO	0.3285	0.3736	0.0141	0.0318	0.1943	0.1743	0.1196

Source: Authors' calculations. Note: We provide the average normalized values for each year between 2018 and 2021

Table 4 Entropy value (e_j) and entropy weight (w_j) for Saudi logistics companies in the period 2018–2021

Entropy values	CR	QR	DER	DAR	AT	ROA	ROE
2021							
e_j	0.9120	0.9119	0.8712	0.9643	0.8925	−0.4259	−0.3213
w_j	0.0275	0.0276	0.0403	0.0112	0.0336	**0.4463**	**0.4135**
2020							
e_j	0.9352	0.9329	0.8899	0.9695	0.9494	−1.1158	−0.9757
w_j	0.0147	0.0152	0.0249	0.0069	0.0115	**0.4793**	**0.4475**
2019							
e_j	0.8563	0.8343	0.8424	0.9474	0.9355	0.7835	0.6022
w_j	0.1199	0.1382	0.1315	0.0439	0.0539	**0.1807**	**0.3320**
2018							
e_j	0.8479	0.8243	0.7661	0.9330	0.9341	0.8687	0.8982
w_j	0.1639	0.1894	**0.2521**	0.0722	0.0711	0.1415	0.1098
2018–2021							
e_j	0.8879	0.8759	0.8424	0.9536	0.9279	0.0276	0.0509
w_j	0.0815	0.0926	0.1122	0.0336	0.0425	**0.3120**	**0.3257**

Source: Authors' calculations by applying the entropy method

Table 5 Ranking of Saudi logistics companies' financial ratios between 2018 and 2021

Financial ratios	Information entropy (e_j)	Entropy weight (w_j)	Ranking
CR	0.8879	0.0815	5
QR	0.8759	0.0926	4
DER	0.8424	0.1122	3
DAR	0.9536	0.0336	7
AT	0.9279	0.0425	6
ROA	0.0276	0.3120	2
ROE	0.0509	0.3257	1

Source: Authors' calculations by applying the entropy method. Note: We provide the average entropy values for the period 2018–2021, the values of each year are available upon request

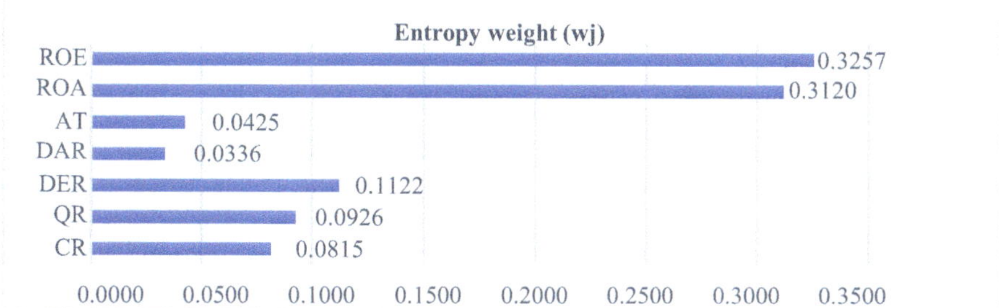

Fig. 1 Entropy weights of Saudi logistics companies' financial ratios between 2018 and 2021

Table 6 Ranking of Saudi logistics companies' financial ratios classified by category between 2018 and 2021

Financial ratio category	Entropy weight (wj)	Ranking
Liquidity ratios	0.1741	2
Solvency ratios	0.1458	3
Efficiency ratios	0.0425	4
Profitability ratios	0.6377	1

Source: Authors' calculations by applying the entropy method. Note: We provide the average entropy values for the period 2018–2021, the values of each year are available upon request

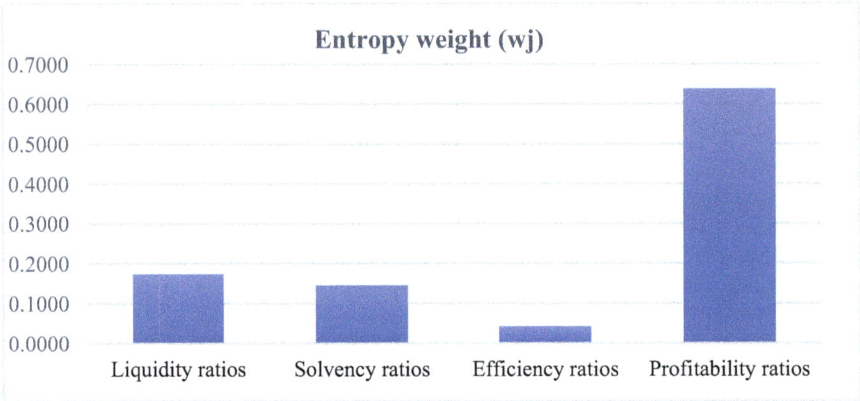

Fig. 2 Entropy weights classified by category of financial ratios in the case of Saudi logistics companies over the period 2018–2021

ratios. The results also show that efficiency ratios (AT) have the least effect on the financial performance of Saudi logistics companies over the period of study.

7 Conclusion

In today's competitive environment, it is clear that improving logistics and supply chain performance has become a major necessity for companies to increase their levels of efficiency and competitiveness and for countries to boost their levels of economic growth [7, 8]. Given the importance of logistics activities in the economy, it is essential that government authorities and managers focus more on maintaining and improving the performance levels in the logistics sector. To achieve this goal, managers of logistics companies and policymakers in the logistics industry need to identify the best performance measures and metrics, especially in a competitive global economy [9, 48]. Although previous literature has focused more on nonfinancial measures, we find that it is also necessary to study the performance of logistics companies from a financial perspective. Indeed, logistics managers need financial measures to make strategic decisions [10]. Despite their importance to decision-makers, the literature reveals that many previous studies have neglected to focus on the financial measures and metrics which are necessary for financial performance measurement of supply chains and logistics companies [29, 35, 44].

It is therefore clear that financial performance measurement of supply chains and logistics companies needs to be further explored in the literature. In this study, we contribute to the literature by investigating the best financial performance measures in the case of ten selected Saudi logistics companies publicly traded on the Saudi Stock Exchange (Tadawul) over the period 2018–2021. The data used in this study include seven financial ratios: current ratio (CR) and quick ratio (QR), which are liquidity ratios; debt-to-equity ratio (DER) and debt-to-assets ratio (DAR), which are solvency ratios; asset turnover (AT) ratio, which is an efficiency ratio; and return on asset (ROA) and return on equity (ROE), which are profitability ratios.

To identify the best financial performance ratios for the selected sample of Saudi logistics companies over the period 2018–2021, we used the entropy method. This method is a multi-criteria decision-making (MCDM) method widely used to select the best performance criteria from different alternatives [11, 48]. The empirical findings indicate that the profitability ratios (ROA and ROE) are the best financial performance evaluation criteria for Saudi logistics companies over the period of study, 2018–2021. The liquidity ratios (CR and QR) and the solvency ratios (DER and DAR) are identified as the performance measures that moderately affect the financial performance of Saudi logistics companies between 2018 and 2021. The results also show that the efficiency ratios (AT) have the least effect on the financial performance of Saudi logistics companies during the period of study.

We believe that the methodology used in this study will be useful to managers and decision-makers in logistics companies. This is because it allows them to identify the best performance indicators of their companies' financial performance in any given period of time and then to make the appropriate decisions based on a founded scientific method. We recommend that managers and decision-makers in logistics companies and supply chain systems apply this methodology when evaluating and analyzing their performance measurement systems.

Furthermore, we suggest that future research studies include in their analysis other categories of financial ratios. Market-based measures, such as Tobin's Q, economic value added (EVA), and market value added (MVA), may be interesting financial measures to be included in the analysis in order to identify the best financial performance criteria for logistics companies [35, 55–57]. Other financial performance measures such as earnings before interest and taxes (EBIT) and other financial performance ratios such as price-to-earnings ratio (PER) could also be used as decision criteria that could affect the financial performance of logistics companies. Other studies could apply the entropy method to any manufacturing-related decision-making processes if the data are available. We also consider that it is very important to continue the research on this topic and seek to improve this work in future research studies by applying any new advances in multi-criteria decision making (MCDM) techniques in the case of logistics firms. Finally, we recommend that other studies use the entropy method to identify nonfinancial performance measures as well as financial performance measures, because of its comprehensive nature and objectivity [58–61].

References

1. A. Harrison, R.V. Hoek, *Logistics Management and Strategy Competing through the supply chain*, 3rd edn. (Pearson, Financial Times Prentice Hall, Harlow, 2008)
2. M. Christopher, *Logistics and Supply Chain Management*, 4th edn. (Pearson, Financial Times Prentice Hall, Harlow, 2011)
3. J. Mehta, Supply chain management in a global economy. Total. Qual. Manag. Bus. **15**(5–6), 841–848 (2004). https://doi.org/10.1080/14783360410001680279
4. Z. Vidrova, Supply chain management in the aspect of globalization, in *19th International Scientific Conference Globalization and its Socio-Economic Consequences – Sustainability in the Global-Knowledge Economy* (Rajecke Teplice, Slovakia, 2019), SHS Web of Conferences 74, p. 04031 (2020)
5. K. Sharipbekova, Z. Raimbekov, Influence of logistics efficiency on economic growth of the CIS countries. Eur. Res. Stud. J **XXI**(2), 678–690 (2018). https://doi.org/10.35808/ersj/1032
6. S. Sezer, T. Abasiz, The impact of logistics industry on economic growth: An application in OECD countries. Eur. J. Soc. Sci. **5**(1), 11–23 (2017). https://doi.org/10.15604/ejss.2017.05.01.002
7. W.P. Wong, C.F. Tang, The major determinants of logistic performance in a global perspective: Evidence from panel data analysis. Int. J. Logist. Res. **21**(4), 431–443 (2018). https://doi.org/10.1080/13675567.2018.1438377
8. P. Hayaloglu, The Impact of developments in the logistics sector on economic growth: The case of OECD countries. Int. J. Econ. Financ. Issues **5**(2), 523–530 (2015)
9. A. Gunasekaran, B. Kobu, Performance measures and metrics in logistics and supply chain management: A review of recent literature (1995–2004) for research and applications. Int. J. Prod. Res. **45**(12), 2819–2840 (2007). https://doi.org/10.1080/00207540600806513
10. F. Saleheen, M.M. Habib, Z. Hanafi, Supply chain performance measurement model: A literature review. Int. J Sup. Chain. Mgt. **7**(3), 70–78 (2018)
11. H.H. Erdoğan, G. Kırbaç, Financial performance measurement of logistics companies based on entropy and WASPAS methods. J. Bus. Res. Turk **13**(2), 1093–1106 (2021). https://doi.org/10.20491/isarder.2021.1186
12. C.D.J. Waters, *Logistics: An Introduction to Supply Chain Management*, 2nd edn. (Palgrave Macmillan, New York, 2009)
13. M. Christopher, *Logistics and Supply Chain Management: Creating Value-Adding Networks*, 3rd edn. (Prentice Hall, London, 2005)
14. J.L. Maia, A.L. Cerra, Interrelation between supply chain management and logistics: A case study in the Brazilian plant of a multinational automotive company. Revista Gestão Ind. **5**(1), 59–73 (2009). https://doi.org/10.3895/S1808-04482009000100004
15. L. Li, *Managing Supply Chain and Logistics, Competitive Strategy for a Sustainable Future* (World Scientific Publishing Company, Singapore, 2014). https://doi.org/10.1142/9151
16. S. Globerson, G. Wolbrum, Logistics management and supply chain management: A critical evaluation. Int. J. Bus. Econ. Res. **3**(2), 82–88 (2014). https://doi.org/10.11648/j.ijber.20140302.15
17. M. Christopher, *The Strategy of Distribution Management*, 1st edn. (Heinemann, Oxford, 1986)
18. R.D. Shapiro, J.L. Heskett, *Logistics Strategy: Cases and Concepts* (West Publishing, St Paul, 1985)
19. B. Cao, L. Deng, Influencing factors of logistics industry growth efficiency in Yangtze River Economic Belt. Econ. Geogr. **39**(7), 148–157 (2019)
20. Y. Zhang, J. Zhao, Empirical study on the contribution of logistics industry development to economic growth in China. China Storage Transp. **8**, 124–126 (2019)
21. B. Cao, J. Rui, Influence of manufacturing agglomeration on spatial evolution of logistics industry: Taking Jiangsu province as an example. Areal Res. Dev. **38**(4), 44–49 (2019)

22. A. Kumar, A. Zhalnin, L.J. Kim, Beaulieu, Transportation and logistics cluster competitive advantages in the U.S. regions: A cross-sectional and spatio-temporal analysis. Res. Transp. Econ. **61**, 25–36 (2017). https://doi.org/10.1016/j.retrec.2016.07.028
23. Y.T. Mohmand, A. Wang, A. Saeed, The impact of transportation infrastructure on economic growth: Empirical evidence from Pakistan. Transp. Lett. **9**(2), 63–69 (2017). https://doi.org/10.1080/19427867.2016.1165463
24. T. Tsekeris, Domestic transport effects on regional export trade in Greece. Res. Transp. Econ. **61**, 2–14 (2017). https://doi.org/10.1016/j.retrec.2016.08.006
25. S.-Y. Kim, H. Park, H.-M. Koo, D.-K. Ryoo, The effects of the port logistics industry on port city's economy. J. Navig. Port Res. **39**(3), 267–275 (2015). https://doi.org/10.5394/KINPR.2015.39.3.267
26. W. Liu, Dynamic coupling mechanism and empirical study on regional logistics system and economy growth. Econ. Geogr. **31**, 1493–1498 (2011)
27. Oxford Business Group (OBG), *Saudi Arabia Invests in Transport Infrastructure to Improve Connectivity* (2022). https://oxfordbusinessgroup.com/reports/saudi-arabia/2022-report/transport/priority-lane-major-infrastructure-projects-and-investment-under-way-to-help-the-kingdom-reach-its-global-connectivity-goals. Accessed 5 Mar 2023
28. Saudi Arabia-Logistics Capacity Assessments (LCAs), Saudi Arabia Logistics Infrastructure (2022). https://dlca.logcluster.org/display/DLCA/Saudi+Arabia?src=contextnavpagetreemode. Accessed 6 Mar 2023
29. H. Neely, J. Richards, K. Mills, M. Platts, Bourne, Designing performance measures: A structured approach. Int. J. Oper. Prod. **17**(11), 1131–1152 (1997). https://doi.org/10.1108/01443579710177888
30. F.T.S. Chan, H.J. Qi, An innovative performance measurement method for supply chain management. Supply Chain Manag. **8**(3), 209–223 (2003). https://doi.org/10.1108/13598540310484618
31. A. Gunasekaran, C. Patel, R.E. McGaughey, A framework for supply chain performance measurement. Int. J. Prod. Econ. **87**(3), 333–347 (2004). https://doi.org/10.1016/j.ijpe.2003.08.003
32. G.P. Kurien, M.N. Qureshi, Study of performance measurement practices in supply chain management. Int. J. Bus. Manag. Soc. Sci. **2**(4), 19–34 (2011)
33. A. Gunasekaran, C. Patel, E. Tirtiroglu, Performance measures and metrics in a supply chain environment. Int. J. Oper. Prod. Manag. **2**(1/2), 71–87 (2001). https://doi.org/10.1108/01443570110358468
34. M.R. Galankashi, S.A. Helmi, Assessment of hybrid lean-agile (leagile) supply chain strategies. J. Manuf. Technol. Manag. **27**(4), 470–482 (2016). https://doi.org/10.1108/JMTM-08-2015-0069
35. M.R. Galankashi, F.M. Rafiei, Financial performance measurement of supply chains: A review. Int. J. Product. Perform. Manag. **71**(5), 1674–1707 (2021). https://doi.org/10.1108/IJPPM-11-2019-0533
36. B.S. Sahay, Supply chain collaboration: The key to value creation. Work Stud. **52**(2), 76–83 (2003). https://doi.org/10.1108/00438020310462872
37. O. Kherbach, M.L. Mocan, The importance of logistics and supply chain management in the enhancement of Romanian SMEs. Procedia Soc. Behav. Sci. **221**, 405–413 (2016). https://doi.org/10.1016/j.sbspro.2016.05.130
38. R. Kumar, S. Singh, Modgil, Exploring the relationship between ICT, SCM practices and organizational performance in the Agri-Food supply chain. Bench Mark. **27**(3), 1003–1041 (2020). https://doi.org/10.1108/BIJ-11-2019-0500
39. F.T.S. Chan, Performance measurement in a supply chain. Int. J. Adv. Manuf. Technol. **21**(7), 534–548 (2003). https://doi.org/10.1007/s001700300063
40. H. Shepherd, Gunter, Measuring supply chain performance: Current research and future directions. Int. J. Product. Perform. Manag. **55**(3/4), 242–258 (2006). https://doi.org/10.1108/17410400610653219
41. P. Taticchi, P. Garengo, S.S. Nudurupati, F. Tonelli, R. Pasqualino, A review of decision support tools and performance measurement and sustainable supply chain management. Int. J. Prod. Res. **53**(21), 6473–6494 (2015). https://doi.org/10.1080/00207543.2014.939239

42. M.M.H. Saleheen, *What to Know About Supply Chain Management* (Nova Science Publishers, 2022). https://doi.org/10.52305/FBAX6283
43. W.J. Ritchie, R.W. Kolodinsky, Nonprofit organization financial performance measurement: An evaluation of new and existing financial performance measures. Nonprofit Manag. Leadership **13**(4), 367–381 (2003). https://doi.org/10.1002/nml.5
44. M. Friemann, P. Wandfluh, R.A. Schonsleben, Improving the application of financial measures in supply chain management, in *IFIP International Conference on Advances in Production Management Systems (APMS)*, (Springer, Berlin/Heidelberg, 2012), pp. 584–591
45. T. Hernaus, M.P. Bach, V.B. Vuksic, Influence of strategic approach to BPM on financial and non-financial performance. Balt. J. Manag. **7**(4), 376–396 (2012). https://doi.org/10.1108/17465261211272148
46. C.E. Shannon, A mathematical theory of communication. Bell Syst. Tech. J. **27**, 379–423 (1948). https://doi.org/10.1002/j.1538-7305.1948.tb00917.x
47. J. Wu, J. Sun, L. Liang, Y. Zha, Determination of weights for ultimate cross efficiency using Shannon entropy. Expert Syst. Appl. **38**, 5162–5165 (2011). https://doi.org/10.1016/j.eswa.2010.10.046
48. W.H. Lam, W.S. Lam, K.F. Liew, P.F. Lee, Decision analysis on the financial performance of companies using integrated entropy-fuzzy TOPSIS model. Mathematics **11**, 397 (2023). https://doi.org/10.3390/math11020397
49. Y. Zhu, D. Tian, Y. Feng, Effectiveness of entropy weight method in decision-making. Math. Probl. Eng. **2020**, 1–5 (2020). https://doi.org/10.1155/2020/3564835
50. M.Z. Rehman, M.N. Khan, I. Khokhar, Investigating liquidity-profitability relationship: Evidence from companies listed in Saudi stock exchange (Tadawul). J. Appl. Financ. Bank. **5**(3), 1792-6580 (2015)
51. G. Sinha, *Financial Statement Analysis. Eastern Economy* (Prentice Hall of India Private Limited, New York, 2012)
52. T.R. Robinson, E. Henry, W.L. Pirie, M.A. Broihahn, *International Financial Statement Analysis*, 3rd edn. (Wiley, Hoboken, 2015)
53. L. Warrad, R. Al Omari, The impact of activity ratios among industrial sectors' performance: Jordanian case. Res. J. Fin. Account **6**(6), 173–178 (2015)
54. M.A.Y. Almumani, An empirical study on effect of profitability ratios & market value ratios on market capitalization of commercial banks in Jordan. Int. J. Bus. Soc. Sci. **9**(4), 39–45 (2018). https://doi.org/10.30845/ijbss.v9n4p639
55. A.D. Ellinger, A.E. Ellinger, B. Yang, S.W. Howton, The relationship between the learning organization concept and firms' financial performance: An empirical assessment. Hum. Resour. Dev. Q. **13**(1), 5–22 (2002). https://doi.org/10.1002/hrdq.1010
56. E. Hofmann, A. Locker, Value-based performance measurement in supply chains: A case study from the packaging industry. Prod. Plan. Control **20**(1), 68–81 (2009). https://doi.org/10.1080/09537280802685272
57. H.-W. Kang, G.W.Y. Wang, H.-S. Bang, S.-H. Woo, Economic performance and corporate financial management of shipping firms. Int. J. Marit. Econ. **18**, 317–330 (2016). https://doi.org/10.1057/mel.2015.8
58. P. Chen, Effects of normalization on the entropy-based TOPSIS method. Expert Syst. Appl. **136**, 33–41 (2019). https://doi.org/10.1016/j.eswa.2019.06.035
59. Z.X. Wang, D.D. Li, H.H. Zheng, The external performance appraisal of China energy regulation: An empirical study using a TOPSIS method based on entropy weight and Mahalanobis distance. Int. J. Environ. Res. Public Health **15**, 236 (2018). https://doi.org/10.3390/ijerph15020236
60. K. Dong, Yang, Application of the entropy-DEMATEL-VIKOR multicriteria decision-making method in public charging infrastructure. PLoS ONE **16**(10), 1–25 (2021). https://doi.org/10.1371/journal.pone.0258209
61. K.F. Liew, W.S. Lam, W.H. Lam, Financial network analysis on the performance of companies using integrated entropy – DEMATEL – TOPSIS model. Entropy **24**, 1056 (2022). https://doi.org/10.3390/e24081056

Index

I. Alharbi et al. (eds.), *Advances in Computational Logistics and Supply Chain Analytics*, Unsupervised and Semi-Supervised Learning,
https://doi.org/10.1007/978-3-031-50036-7

www.ingramcontent.com/pod-product-compliance
Ingram Content Group UK Ltd.
Pitfield, Milton Keynes, MK11 3LW, UK
UKHW022211051026
15624UKWH00001B/105
* 9 7 8 3 0 3 1 5 0 0 3 5 0 *